# Stochastic Modelling of Drinking Water Treatment in Quantitative Microbial Risk Assessment

*KWR Watercycle Research Institute Series*

# Stochastic Modelling of Drinking Water Treatment in Quantitative Microbial Risk Assessment

Patrick W.M.H. Smeets

Publishing
London • New York

**Published by** **IWA Publishing**
**Alliance House**
**12 Caxton Street**
**London SW1H 0QS, UK**
Telephone: +44 (0)20 7654 5500
Fax: +44 (0)20 7654 5555
Email: publications@iwap.co.uk
Web: www.iwapublishing.com

First published 2011


Cover design by www.designforpublishing.co.uk
Typeset by MPS Limited, a Macmillan Company, Chennai, India, www.macmillansolutions.com



*British Library Cataloguing in Publication Data*
A CIP catalogue record for this book is available from the British Library

*Library of Congress Cataloging-in-Publication Data*
A catalog record for this book is available from the Library of Congress

ISBN13: 9781843393740
ISBN: 1843393743

# Contents



 ISBN: 9781843393740. Published by IWA Publishing, London, UK.

# Acknowledgements

The research for this book was funded by the MicroRisk project and the joint research program of the Dutch water companies. The MicroRisk (www.microrisk.com) project is co-funded by the European Commission under contract number EVK1-CT-2002-00123.

I would like to thank all the people who helped me with the research and the work presented in this book, especially the following persons.
Gertjan Medema
Luuk Rietveld
Hans van Dijk
Alex van der Helm
Pieter van Gelder
Geoff Stanfield
Susan Petterson
Wim Hijnen
Nick Ashbolt
Paul Hunter
Thor-Axel Stenström
Yolanda Dullemont

# Summary

## STOCHASTIC MODELLING OF DRINKING WATER TREATMENT IN QUANTITATIVE MICROBIAL RISK ASSESSMENT

Drinking water outbreaks of infectious disease in the twentieth and twenty-first centuries made clear that the absence of indicator organisms did not guaranty that the drinking water was safe. The World health organization (WHO) has developed the Water Safety Plan (WSP) approach to address the shortcomings of the indicator principle. The WSP aims to achieve safe drinking water by adequate control of drinking water sources, treatment and distribution. The ability of the total system to provide safe drinking water and the required activities to verify that safe water is provided are assessed in the WSP. Since conditions can vary between individual systems, site specific studies are required to assess the safety of a drinking water system. In addition, it has become clear that very short periods of unsafe water can have a major impact on the (mean) risk of infection from drinking water. Quantitative microbial risk assessment (QMRA) was developed to assess the level and variability of the health risk and can include the uncertainty involved in the assessment. It is therefore logical to apply QMRA in the WSP at points where risks need to be quantified.

Health-based targets can be set to determine whether the drinking water is safe enough. Commonly applied health-based targets are the maximum acceptable concentration of pathogenic microorganisms in drinking water, the risk of infection and disability adjusted life years (DALY). The health-based target in Dutch drinking water legislation is a maximum of 1 infection per 10,000 people per year. This is roughly similar to a concentration of one virus per one million litres or $10^{-6}$ DALY. In QMRA the risk of infection is calculated from the number of pathogenic microorganisms a person is exposed to (the dose) and the chance that this person develops an infection (the dose-response). Microbial analysis of one million litres of water is not feasible. In QMRA the number of pathogenic microorganisms in drinking water can be calculated from their concentration in the source, for example surface water and their reduction by drinking water treatment. Previous research indicated that one of the largest uncertainties in QMRA comes from estimating the efficacy of drinking water treatment. Goal of the current study presented in this book was to improve the quantification of the efficacy of drinking water treatment for the reduction of (pathogenic) microorganisms.

 ISBN: 9781843393740. Published by IWA Publishing, London, UK.

In the WSP, several legislations and industry standards, treatment efficacy is determined through the application of 'log credits'. For example, through experiments it was found that filtration removed 99% of a bacteria species, which equals two log credits (bacteria are removed by two orders of magnitude). Generally a 'conservative mean value' is chosen based on a literature survey of similar experiments. However, the results from these studies can vary over several orders of magnitude. These various observations are a cause of uncertainty with regard to the treatment efficacy of a specific treatment system. In the current work, uncertainty was studied by stochastically modelling treatment efficacy. The efficacy was described by a triangular probability density function (PDF) in the stochastic treatment model. The parameters of the PDF were chosen so that they resembled the various efficacies reported in the literature. In the example, *Cryptosporidium* removal by filtration varied from 0 to 5 log units, with most studies reporting 2 log removal. From the stochastic model it became clear that, based on this knowledge, only 1.5 log removal was expected, and there was a 5% chance that removal was even lower that 0.8 log units. On the other hand there was a 50% chance that removal exceeded 2 log units. For a system that requires several log units of *Cryptosporidium* reduction it may well be worthwhile to assess the achieved site specific removal.

For some processes, such as chemical disinfection, there is a clear relationship between the exposure of microorganisms to the disinfectant and the inactivation of microorganisms. Therefore, in the WSP, the efficacy of these processes is generally calculated with $CT$ models, in which $C$ is the concentration of the disinfectant and $T$ is the contact time. Several to many log credits for inactivation are often awarded based on these calculations. The current study showed that these models do not take the limitations of full-scale treatment systems into account. Due to hydraulic shortcomings and practical limitations, the mixing of the disinfectant and contact time at full-scale often differed from the conditions during experiments from which the $CT$ models were developed. The actual efficacy at full-scale is often limited to 2 to 3 log units. The research showed that at an ozonation system where over 6 log units of inactivation was expected only 2 log units of inactivation was achieved at full-scale.

Besides the uncertainty with respect to the full-scale hydraulic situation, the susceptibility of the microorganism to the disinfectant was a source of uncertainty. This susceptibility is generally investigated using freshly cultured organisms in a laboratory setting resulting in inactivation rate constants that describe inactivation kinetics. The study showed that environmental populations of microorganisms were more resistant to the disinfectant than cultured populations. Cultured microorganisms that survived temporal exposure to environmental conditions were also more resistant than freshly cultured microorganisms. The application of very conservative values for the inactivation rate constants is therefore recommended when modelling full-scale systems.

Since the source water quality and the efficacy of drinking water treatment can vary by orders of magnitude, the site-specific situation needs to be assessed. The characteristics of the source water, treatment process conditions, process monitoring and microbial monitoring of the water were combined in the treatment assessment framework to assess the level and variation of treatment efficacy. Various types of information were used to highlight the elements of the assessment.

Extensive microbial monitoring of drinking water provided insight in the way microorganisms were distributed in drinking water. In the study the results of the daily, continuous monitoring of *Cryptosporidium* in the UK were analysed. This showed that the *Cryptosporidium* concentration generally followed a continuous curve of regular low concentrations and rare high concentrations. Since extremely high concentrations rarely occurred, these were generally regarded as an 'event'. However, these events were the result of normal variations of the system and were therefore referred to as 'normal events'. The frequency and magnitude of these events can be predicted from observations of regular variations through statistical analysis. These predicted events need to be included in QMRA since the arithmetic mean concentration, and thus the mean risk, is dominated by these normal events. In approximately 30% of the 216 studied systems a curve break was observed where extremely high concentrations of *Cryptosporidium* occurred unexpectedly. In these cases a 'special event' occurred, such as a treatment failure, an operational error or an unusual peak contamination of the source water. The frequency and magnitude of 'special events' cannot be predicted by statistical analysis of observations of regular variation of treatment efficacy. However, very frequent observations such as on-line monitoring of process conditions may detect such special events. None or only a few *Cryptosporidium* were detected in the 450 samples of 1,000 litre at systems with high drinking water quality. At these systems it is likely that nominal concentrations, below detection limit, dominate the arithmetic mean *Cryptosporidium* concentration in drinking water. The developed QMRA methods to calculate the distribution of pathogens in drinking water from the concentration in the source water and the reduction by treatment can be applied at these systems to calculate the mean concentration.

The uncertainty involved in microbial analysis methods needs to be included when microbial monitoring data is used in risk assessment. The variable recovery and indirect quantification methods (e.g. presence-absence tests) are examples of uncertainties that are introduced by microbial methods. In addition the way microorganisms are distributed in water may be unknown and the concentration of microorganisms varies in time. Methods to separately quantify uncertainty and variation were developed in the study. By plotting the monitoring data and the data analysis results as a complementary cumulative distribution function (CCDF) the focus of the graph was put on the rare events of high concentrations. Since these high concentrations dominate the mean concentration, accurate assessment of these high concentrations is essential. From the graph, the concentrations that dominate the mean concentration were determined. In most cases the concentrations that occurred 1% to 5% of the time dominated the mean concentration. However, in some cases special events that occurred only 0.1% of the time dominated the mean concentration.

Microbial data from regular monitoring programmes was collected from water supply companies. The data was analysed to assess site specific treatment efficacy. Similar studies reported in literature compared samples before and after treatment taken on the same day to assess treatment efficacy. The current study showed that this led to an overestimation of the variability of treatment, which consequently led to underestimation of treatment efficacy. Therefore an improved method was developed to calibrate the stochastic treatment model with microbial monitoring data. With the improved method, the predicted concentrations after treatment corresponded to the monitored concentrations. Model calibration provided information on treatment performance and was thus used for treatment assessment. The predictive value of the calibrated model was tested by splitting the datasets for calibration and validation. At full-scale, generally only indicator organisms are monitored. In studies where both indicator and pathogenic microorganisms were monitored, there appeared to be little correlation between their removals when comparing samples from the same day. However, calibrating the stochastic model with indicator-organism data did provide an effective model of pathogen reduction. Apparently the variation of treatment efficacy for indicators and pathogens was similar, but this was obscured in daily observations due to the over dispersed distribution of microorganisms in water and temporal variations.

These applications of the stochastic model all aimed to assess the ability of the system to provide water that complies with the health-based target. This is also the first step in the WSP. The current WSP manual applies semi-quantitative methods to estimate the potential health effect and applies log credits and *CT* models to estimate treatment efficacy. QMRA methods can improve the quantification and include the uncertainty of these assessments. Stochastic QMRA can be applied to predict the frequency and magnitude of normal events to estimate the mean risk more accurately. The uncertainty of the model outcome can be assessed by stochastic modelling of variables and parameters in process models. By using site specific information such as on-line disinfectant residual measurements, the results of process variations and control can be included in the risk assessment. Consequently the effect of process improvements can be estimated beforehand to support decisions by risk managers. Examples showed that doubling disinfectant dose had little effect on the efficacy of ozone and chlorine disinfection. However, improving hydraulic conditions and process control could double the efficacy of these processes without the need for more disinfectant.

The frequency and magnitude of a special event cannot be predicted by QMRA. However, when a special event is identified in a WSP, QMRA can be used to quantify the health effect of such an event. Thus risks from various events can be compared and prioritised and monitoring can be designed to detect such a special event. The required monitoring to detect relevant events is directly related to the efficacy of the treatment process. Weekly monitoring is sufficient for 1 log reduction, and daily monitoring is sufficient for 2 log reduction. However, a monitoring frequency of ten seconds is required when 6 log reduction is the treatment goal since even very short moments of failure will affect the mean reduction. It is therefore easier to monitor multiple barriers with limited efficacy than a single, very effective barrier. This type of monitoring needs to verify that the process is running within specifications, for example by monitoring turbidity or disinfectant residual on-line.

In theory a treatment system can be operated such that the required efficacy is exactly achieved. However, a full-scale treatment system is not a large laboratory; therefore process control needs to take into account the variations of processes and equipment, the efficacy of corrective actions and the required response time. QMRA can be used to determine setpoints and critical limits in the WSP such that, even when events occur, the water will continue to comply with the health-based target without excessive costs or other disadvantages. Assessing the ability to meet health-based targets, determining setpoints and critical limits, designing microbial and on-line monitoring and preparing corrective actions are all examples of QMRA applications in a WSP.

The study has provided several scientific methods and techniques that can be applied directly in drinking water practice. Implementing these methods will require investment of resources. However, this investment is far less than the costs resulting from overestimation of the risk which could lead to unnecessary expansion of treatment, or costs following an outbreak when the risk is underestimated. Some elements of QMRA are outside the scope and expertise of the drinking water industry. The legislator (or a central drinking water organisation) could support the water companies by providing guidelines for the acceptable uncertainty of the assessed risk, the applicable dose-response relations and the choice of index pathogens. Thus the drinking water industry can now go beyond the indicator concept, by using QMRA to support proactive management that can ensure the provision of safe drinking water.

# Chapter 1

# Introduction

P.W.M.H. Smeets[a,b], L.C. Rietveld[b], J.C. van Dijk[b], G.J. Medema[a,b], W.A.M. Hijnen[a] and T.A. Stenström[c]

[a]KWR Watercycle research institute, PO Box 1072, 3430 BB Nieuwegein, the Netherlands
[b]Faculty of Civil Engineering, Delft University of Technology, PO Box 5048, 2600 GA Delft, the Netherlands
[c]Department of Parasitology, Mycology, Water and Environmental Microbiology, Swedish Institute for Infectious Disease Control, SE-171 82 Solna, Sweden

## HISTORY OF MICROBIALLY SAFE DRINKING WATER

From the beginning of time man has learned to carefully choose drinking water in order to reduce the risk of illness. Drinking water supply started with the rise of civilisations. The population in cities and communities needed to be provided with safe drinking water, while in the mean time water was increasingly polluted by the communities. This led to waterborne outbreaks of infectious disease, which were already recorded by the Egyptians in 3180 BC (Rose and Masago, 2007). Outbreaks continued to occur through the ages, as the relationship between faecal pollution of the water and outbreaks had not been recognised.

Drinking water treatment of surface water was originally started to improve the aesthetic properties of drinking water. By the time of the Egyptians (15th–13th century BC) and Romans (300 BC–200 AC) settling was applied to reduce turbidity and in the 5th century B.C. Hippocrates, the Father of Medicine, invented the 'Hippocrates Sleeve', a cloth bag to strain rainwater. Supply of settled and filtered water in modern times started in 1804 (Scotland) and 1806 (Paris). Initially slow sand filters were used to provide a more aesthetic product and soon filtration was recognised to reduce outbreaks of typhoid and cholera. In the 1870's Robert

 ISBN: 9781843393740. Published by IWA Publishing, London, UK.

Koch studied water filtration systems that were effective in removal of bacteria after the Hamburg cholera outbreak of 1892. In his biography of Koch's work, Brock (1988) states that 'water filtration has probably saved more lives than immunization and chemotherapy combined'. In 1906 the first ozonation plant for disinfection was started in France. John Snow already promoted chlorination after his pioneering epidemiologic studies during London's cholera outbreaks of the 1850's. Still chlorination became common practice only around 1910. From 1920 the combination of sedimentation, filtration and chlorination virtually eliminated epidemics of classical waterborne diseases, such as cholera and typhoid, in areas so supplied (AWWA, 2006). However, outbreaks of waterborne disease due to poor drinking water quality still occur today, even when treatment is in place. From 1974 to 2002, 26 out of 35 outbreaks in the USA and Canada, as reported by Hrudey and Hrudey (2004), were due to surface water treatment failure or inadequate treatment to deal with sudden peak increases of pathogen concentrations in source water. Some major outbreaks like that of cryptosporidiosis in Milwaukee where treatment efficiency was compromised would have been prevented or the impact on human health reduced, by adequate treatment. So despite modern water treatment, means to verify that the water is safe to drink are still required.

By the end of the nineteenth century, the presence of specific bacteria in drinking water was recognized as an indicator of unsafe water. The use of coliforms as indicator organisms to judge the microbial safety of drinking water was initiated (Greenwood and Yule, 1917). The absence of indicator organisms such as *Escherichia coli* in drinking water is still part of most legislation today. In the 1970's the shortcomings of coliforms became clear. Newly recognized waterborne pathogens, such as viruses and protozoa turned out to be more resistant to drinking water treatment processes such as chlorination than coliforms. The search for other, more resistant indicator organisms such as bacterial spores and bacteriophages was started. Their applicability turned out to be limited, as outbreaks continued to occur even when no indicator organisms were detected (Hrudey and Hrudey, 2004). Large drinking water related outbreaks were generally picked up by epidemiology, but the prevalence of endemic illness caused by drinking water was so low in most developed countries that epidemiology was not sensitive enough to identify the source (Taubes, 1995). Apart from monitoring drinking water for the absence of indicator organisms, other ways to protect the drinking water consumer were sought. In the 1970's the National Academy of Sciences initiated chemical risk assessment for drinking water resulting in the 'Safe drinking water act' in 1974 (SDWA, 1974). Analogous to the chemical risk targets, a target for risk of infection (not illness) below $10^{-4}$ per person per year was being advocated in the USA.

Between 1983 and 1991 quantitative microbial risk assessment (QMRA) was used sporadically to assess microbial risks in drinking water (Haas, 1983; Gerba and Haas, 1988; Regli *et al.*, 1991 and Rose *et al.*, 1991). These first assessments were focussed on producing a reliable dose-response relationship for the very low pathogen doses expected in drinking water. These led to the 'single hit theory' stating that exposure to a single pathogenic organism could lead to infection and subsequently illness. The studies calculated the risk of infection from the monitored or estimated pathogen concentrations in drinking water. These studies recognised the limitations of drinking water monitoring for QMRA. Regli *et al.* (1991) concluded that: '*Inordinately large numbers of high-volume samples (generally a total volume of >100,000 to 1,000,000 L) are required to ascertain whether a potable water is below the $10^{-4}$ risk level. Thus, finished-water monitoring is only practical to determine whether a very high level of risk exists, not whether a supply is reasonably safe.*' Hrudey and Hrudey (2004) showed that the occurrence of false positives makes it virtually impossible to estimate indicator bacteria concentrations in drinking water by monitoring at the observed low level. However, direct monitoring of pathogens in drinking water has been applied. The statutory *Cryptosporidium* monitoring (DWI, 1999) in the UK has been the most extensive monitoring program for pathogens in drinking water and is further discussed in Chapter 3.

To overcome the shortcomings of drinking water monitoring, computational methods were applied in QMRA. Regli *et al.* (1991) stated that: '*Determining pathogen concentration (or demonstrating its absence) in source waters and estimating the percentage-removal or inactivation by treatment allow for risk estimates of pathogen occurrence in finished water and the associated risk of infection.*' Subsequent studies found that quantifying treatment efficacy introduced substantial uncertainty in QMRA (Teunis *et al.*, 1997; Gibson III *et al.*, 1999 and Payment *et al.*, 2000). From the outbreaks it had become clear that short hazardous events could have a significant impact on public health. In addition, the financial consequences of an outbreak may well make these events important to identify and advert (Signor and Ashbolt, 2007). Although counteracting peak events is necessary to prevent outbreaks, sufficient treatment during baseline (normal) conditions is also required to achieve an acceptable level of endemic infections. In specific situations the sporadic cases (during baseline conditions) appeared to represent a greater proportion of waterborne disease than outbreaks (Nichols, 2003). This was also a conclusion reached for a water supply system in Gothenburg, based on failure reporting and QMRA (Westrell *et al.*, 2003).

## STATE OF THE ART OF QMRA IN 2002

### Treatment assessment for QMRA

Regli *et al.* (1991) first suggested monitoring pathogens in source water and modelling the removal by treatment. Initially rules of thumb and engineering guidelines were used to provide a point estimate of treatment efficacy. Rose *et al.* (1991) used QMRA to determine the required treatment efficacy to reach health-based targets, rather than actually assess the efficacy. As more research was performed, it became clear that treatment efficacy could vary substantially between treatment sites. LeChevallier *et al.* (1991) assessed treatment efficacy for *Cryptosporidium* and found substantial differences in treatment efficacy at very similar sites. These could not be explained by treatment characteristics such as filter to waste practice or choice of coagulant. Payment *et al.* (1993) studied removal and inactivation of viruses and indicator organisms. He used the mean of the observed concentrations before and after treatment steps to quantify treatment efficacy, thus disregarding the effect of treatment variations. Other QMRA studies did not model treatment but started from a concentration in treated water, such as Haas *et al.* (1993) who based virus concentrations in drinking water on Payment (1985). Similarly Crabtree *et al.* (1997) did not estimate treatment efficacy for virus removal but assumed concentrations in drinking water of 1/1000 and 1/100 virus per litre. Gerba *et al.* (1996) assumed 4 log reduction of rotavirus by treatment based on the Surface Water Treatment Rule (SWTR, USEPA, 1989) credits. Teunis *et al.* (1997) incorporated the variation in time and the uncertainty with regard to the efficacy at a specific site in a stochastic QMRA by the use of PDFs to describe the concentrations of microorganisms and treatment efficacy. Microbial monitoring data before and after treatment were paired by date to provide a set of reduction values, and the PDF was fitted to these. Their conclusion was that (variation of) reduction by treatment dominated the uncertainty of this risk. Haas *et al.* (1999) provided an overview of methods for QMRA both in drinking water and other fields such as recreational waters and food. They found that identification of distributional form may be subject to error if a limited amount of data points are used. Consequently the risk analysis should not put too much weight on the tails of these distributions which would represent rare events of poor treatment. Haas *et al.* (1999) also discussed the use of monitoring data (virus removal by lime treatment) and process models (for virus decay in groundwater and chemical inactivation) to assess treatment efficacy. Teunis and Havelaar (1999) performed a full QMRA, including quantification of treatment efficacy using monitored reduction of Spores of Sulphite-reducing Clostridia (SSRC) as a surrogate for *Cryptosporidium* removal. Variability of filtration was modelled by a two-phase model: 'good removal' and 'poor removal'. Medema *et al.* (1999) applied similar methods. Variability of ozonation was modelled by running an inactivation model with monitored ozone concentrations. Payment *et al.* (2000) used log credits from the SWTR in risk assessment of *Giardia* since '*Attempting to actually enumerate indicator microorganisms or pathogens under actual plant conditions rarely provides useful data*'. Dewettinck *et al.* (2001) assessed the safety of drinking water production from municipal wastewater based on treatment efficacy reported in literature. Fewtrel *et al.* (2001) assessed the uncertainties in drinking water QMRA and found that treatment contributed the least uncertainty. However, this was based on a single experiment of *Cryptosporidium* removal by treatment. A 2001 USEPA study on *Cryptosporidium* removal (USEPA, 2001) found large ranges of removal (typically over 3 log) and generally less removal at full-scale than at laboratory or pilot scale. In an extensive literature review of treatment efficacy by LeChevallier and Au (2004), large variations in treatment efficacy between studies was found. Masago *et al.* (2002) applied QMRA to assess the risk from *Cryptosporidium*, including the effect of rare events. Treatment was modelled bimodally with good removal (99.96%) or poor removal (70.6%).

In general it could be concluded that most QMRA studies used log credits to model treatment performance, which were not site specific. Site specific assessment of treatment efficacy for QMRA indicated that treatment efficacy at full-scale could be significantly higher or lower than the applied log credits (Teunis *et al.*, 1997, 1999; Teunis and Havelaar, 1999 and Medema *et al.*, 1999). Moreover, such an assessment could provide management strategies to be applied at the site to improve drinking water safety. Site specific assessment was complicated by the way pathogens were distributed in water, treatment variations and correlation between treatment steps.

## Distribution of pathogens in water

Pipes *et al.* (1977) found that organism counts in 100 ml samples from a 10 L sample were not necessarily Poisson distributed, which would be expected if the organisms were randomly dispersed in the water. Gale *et al.* (1997) found that although *Bacillus subtilus* var. *niger* were Poisson distributed in raw water, this was not the case in treated water (within a 500 ml sample). He concluded that treatment could change the distribution of microorganisms in the water. As a consequence, in an outbreak overdispersion would lead to some individuals ingesting high numbers of pathogens and some not receiving any. In combination with the dose-response relationship, this might have an impact on the assessed risk. At low doses the risk of infection would be determined by the arithmetic mean concentration. The arithmetic mean is dominated by the rare high concentrations when organisms are over-dispersed. Quantifying these high concentrations is problematic due to their rarity. Gale (2001) also showed that organisms are not completely dispersed in drinking water. The (lack of) relation between influent and effluent samples observed by Teunis *et al.* (2004) might partly be caused by the over dispersion of microorganisms. The change of distribution of microorganisms in water due to water treatment processes was likely to affect the observed reduction by treatment from microbial monitoring.

## Treatment variation and rare events

From stochastic QMRA studies it became clear that when variations were incorporated, rare events of high pathogen concentrations or poor treatment could dominate the risk of infection. Haas and Trussell (1998) compared a system redundancy method to a stochastic method as a way of incorporating rare events of poor treatment. The system redundancy method was based on log credits per treatment step. Compliance of reduction by the total treatment was required even when one barrier failed completely (rare event). The stochastic method applied a probability density function (PDF) of likely performance to the separate barriers and combined these in a Monte Carlo simulation to predict total treatment efficacy for QMRA. The importance of good PDF fit for very skewed data was stressed, implying that high numbers of data points were required. Gibson III *et al.* (1999) identified exposure assessment (including treatment assessment) as one of the important fields of research for risk assessment of waterborne protozoa due to the uncertainty about and variability of protozoan reduction by treatment. Teunis *et al.* (1999, 2004) explored various methods to quantify variation of treatment efficacy. They found that the extremes of the distributions of treatment efficacy (and other factors such as recovery) dominated the assessed risk. The approach of statistical analysis of fractions was more appropriate than often used calculations based on the ratio between the (geometric) means 'before' and 'after' treatment. Masago *et al.* (2002) found that eliminating rare occurrences (<1% of time) of high concentrations exceeding 1/80 L was required to reduce the risk to $10^{-4}$ per person per year. This demonstrated the impact of rare events on average risk and the need to estimate the frequency and magnitude of rare events of poor treatment in QMRA.

## Correlation between treatment steps

Initially the efficacy of all steps in treatment was considered to be independent (Regli *et al.*, 1991). Smith *et al.* (1992) stated that potential dependencies between parameters in QMRA needed to be addressed. However, when correlations between 'relatively well known variables' were concerned, analysts should focus on better quantifying the key factors rather than to focus on correlations. Bukowski *et al.* (1995) showed that the choice of PDF type could dominate the risk assessed through Monte Carlo simulation. The impact of correlations was only significant when correlations were very high. Medema *et al.* (1999) suggested that interaction between treatment processes might cause correlation between the efficacies of consecutive treatment steps. They suggested this correlation could cause the deviance between predicted and monitored concentrations after treatment in their QMRA study. Haas (1999) explored the use of copulas to describe correlations between the random variables in Monte Carlo simulation. He concluded that the chosen form of correlation may have a significant impact on the results. He did not specify how to determine the correlation within a QMRA of drinking water. Correlation was not incorporated in stochastic QMRA studies by Teunis *et al.* (1997, 1999). More recently a correlation between concentration of microorganisms and reduction efficacy was suggested by Haas and Kaymak (2003). So far the studies have been inconclusive on the occurrence of correlation between treatment steps and the need to incorporate correlation in QMRA.

## Direct assessment of pathogens in drinking water

Apart from modelling treatment, direct assessment of pathogens in drinking water was also performed. Isaac-Renton *et al.* (1999) tried to correlate *Cryptosporidium* levels in drinking water to seroprevalence of antibodies in three communities. None of the supplies were filtered and only chlorination was applied. No *Cryptosporidium* was found in deep-well water, and only few in water from a well protected catchment, whereas 20% of samples in water from the unprotected watershed were positive for *Cryptosporidium*. However, there were no significant differences in seroprevalence rates, which ranged from 33% to 53%. No direct link between monitored drinking water quality and infection of the population could be established. Lloyd and Drury (2002) evaluated the first results of the UK statutory *Cryptosporidium* monitoring. After nine months of sampling, 0.23% of the samples were non-compliant (i.e. contained more than 1 oocysts per 10 L) and no outbreaks had been associated with these observations. Oocysts were detected in 8.9% of all samples, covering 44% of the sampled sites. Even though most of these samples were in compliance, levels of 0.01 to 0.10 oocysts per 10 L could pose a health risk. Hellard *et al.* (2001) went one step further and investigated the effect of microbial water quality on rates of community gastroenteritis in Melbourne by measuring the difference in the levels of illness among two population groups, each comprising approximately 300 households. One group consumed normal tap water and the other consumed water that was filtered and disinfected with ultraviolet radiation. The study found no measurable difference in illness rates between the normal tap water group and the filtered water group, thus demonstrating that Melbourne's unfiltered drinking water does not make a significant contribution to gastroenteritis rates (8% being the detection limit). These studies showed the limitations of drinking water monitoring and epidemiology to assess low levels of risk at reasonable costs.

## QMRA in drinking water guidelines and legislation

After the first attempts of Quantitative Microbial Risk Assessment (QMRA) in 1983, QMRA was applied in various ways to improve the microbial safety of drinking water. In 1996 the ILSI Risk Science Institute Pathogen Risk Assessment Working Group developed a conceptual framework to assess the risks of human disease associated with exposure to pathogenic microorganisms (ILSI, 1996) which was based on QMRA. This was later evaluated by Teunis and Havelaar (1999). Haas *et al.* (1999) wrote an extensive guide to risk assessment for pathogens in (drinking) water.

In 1989 the USEPA used QMRA to develop technical requirements for drinking water treatment in the Surface Water Treatment Rule (SWTR, USEPA, 1989) in order to roughly achieve a maximum risk of infection of $10^{-4}$ per person per year for *Giardia* and viruses. Later the SWTR was extended for *Cryptosporidium* in the IESWRT (USEPA, 1998), and was further elaborated in the LT1ESWTR (2002) and LT2ESWTR (2006). The rule awards 'reduction credits' for treatment processes when these are sufficiently monitored. The combined processes need to provide sufficient treatment for the level of source water contamination.

Other regulators did not set technical standards; instead they required a site specific QMRA for each drinking water system. In 2001 Dutch drinking water regulations included a maximum acceptable risk of infection of $10^{-4}$ per person per year, to be verified with QMRA (Anonymous, 2001). The WHO Water Quality: Guidelines, Standards and Health (WHO, 2001) presented a harmonized framework for risk assessment and management. Apart from risk of infection, WHO promoted a risk endpoint of $10^{-6}$ disability adjusted life years (DALY) which includes the adverse health effects when an infected individual becomes ill. The reader is referred to (WHO, 2004) for a complete explanation of the DALY. The new proposed Canadian drinking water guidelines for viruses include QMRA to verify that sufficient treatment is applied to reach a health-based target of $10^{-6}$ DALY (CDW, 2007). QMRA was also considered for legislation of bathing water (USEPA, 2007) and in Australia for water reuse (NWQM, 2006, 2007).

A third development to improve drinking water safety focussed on managing risks on an operational level. In 1994 the use of Hazard Analysis and Critical Control Point (HACCP), as applied for food safety, was tested for applicability in drinking water safety (Havelaar, 1994 and Teunis *et al.*, 1994). Over the years this concept developed into Water Safety Plans (WSP) (Barry *et al.*, 1998; Deere and Davison, 1998 and Davison *et al.*, 2006). In 2004 the IWA and WHO presented the Bonn charter (IWA/WHO, 2004) which set a high level framework for drinking water risk management. In addition WHO published the third edition of the Drinking Water Guidelines (WHO, 2004). Both promoted the use of Water Safety Plans (WSP) to manage drinking water safety in an integral manner. In 2002 the MicroRisk project was started (MicroRisk, 2002 and Medema *et al.*, 2006) to bring together the WSP and QMRA methods.

## QMRA: ITS VALUE FOR RISK MANAGEMENT

At various steps in the HACCP-based process of the water safety plan (WSP), questions emerge that relate to the balance between safety and costs of the water supply system. More safety can be obtained by including additional control measures, by setting very strict limits, by intensive monitoring etc. However, resources are not unlimited and drinking water is not the only transmission route for pathogens and toxic compounds that need to be controlled. QMRA provides information for efficient allocation of resources to water supply. By setting health-based targets based on the contribution of drinking water to the overall health risk of the human population, it becomes clear when *safe* is *safe enough*. Links between QMRA and WSP are illustrated by the questions it answers in Figure 1.1. Most of these questions especially relate to drinking water treatment, since it is there that the (polluted) source water is transformed to safe drinking water.

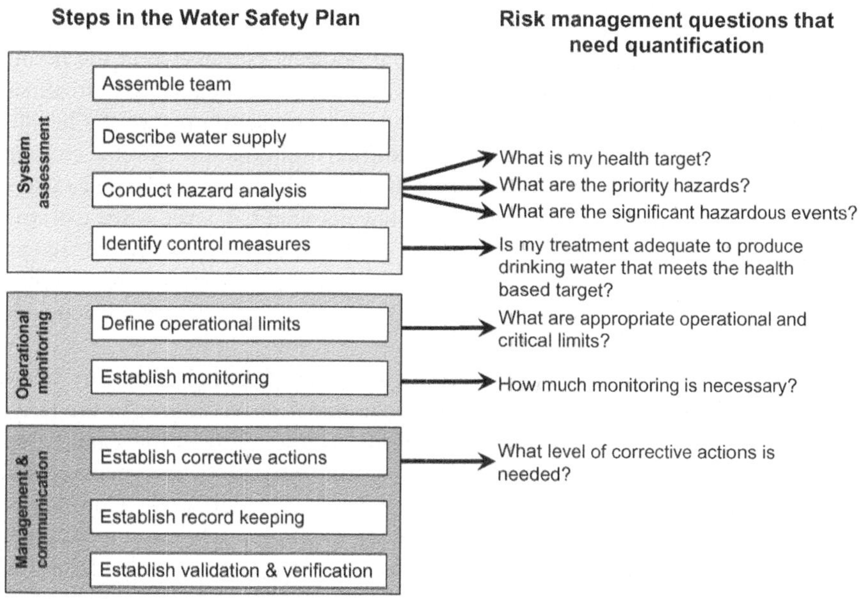


**Figure 1.1** Risk management questions that can be quantified by QMRA (from: Medema *et al.*, 2006)

## Complying with health targets

At the water utility level, a QMRA can be conducted to answer the question: 'Do we meet the health target?'. It is the responsibility of the water utilities to meet the health-based targets and to demonstrate to the regulators and the public that these targets are met. During the WSP process the risks are approached in a semi-quantitative manner (high, medium, low etc.), based on experience, industry standards and subject to personal interpretation. In many cases, this is sufficient information for risk management; i.e. it is clear that a well-head that is not properly closed may give rise to contamination of the water from the well and the corrective action will be to close the well-head properly. In these cases, there is usually no further quantitative assessment of the risk of contamination necessary to trigger the appropriate corrective actions. However, this does not answer the question whether the overall water supply system from source-to-tap provides safe drinking water to the consumer. A quantitative microbial risk assessment of a drinking water system can demonstrate that the health-based targets are met.

A QMRA (in the WSP: System assessment) is therefore the logical first step when safety of a water supply system is under consideration. QMRA cannot only provide a quantitative estimate of the level and variation of risk. It also provides an indication of the uncertainty of the assessment, allowing for a balanced interpretation of the outcome. If the outcome of the assessment indicates that the drinking water could be unsafe under some conditions, QMRA can help to identify the most economic, sufficiently effective measure to bring the risk within the health-based targets. When drinking water is produced from surface water, drinking water treatment generally forms the means by which the water quality is controlled. Since direct assessment of drinking water safety through drinking water pathogen monitoring is not feasible, quantifying treatment efficacy is a crucial step in QMRA.

## Quantifying normal events and special events

Bartram *et al.* (in WHO, 2001) identified that QMRA should not only be directed at the nominal performance of treatment systems, but also at the moments of poor source water quality and treatment performance. These moments, referred to as hazardous events in the WSP, may comprise most of the health risk. This book distinguishes between 'normal events' and 'special events'. Normal events are the extreme consequence of normal variations in the system, such as seasonal variations of temperature, filter backwash cycles and chemical dosing control. Although these variations normally balance out to a low nominal risk level, some extreme combination of conditions can lead to an event. These normal events can be predicted by extrapolating normal variations, similar to extrapolating wind velocities to predict the one in 1,000 year storm. Whereas normal events may come as a surprise to risk managers faced with an extreme variation, QMRA can estimate the frequency and magnitude of normal events based on observed nominal variation.

Special events are not part of these normal variations. Examples of special events are treatment equipment failure or human error, which cannot be predicted based on nominal observations. In the HACCP-based WSP system, special events in treatment were identified and prioritized through fault trees and Risk Factor Matrices (Davison *et al.*, 2006). These methods relied on experience and insights of risk managers and operators to identify events and quantify the actual effect of an event on drinking water safety. However, the effect of e.g. a dosing pump failure on health is hard to quantify intuitively. QMRA can be used to quantify the effect of a special event on consumer's health. This allows the risk manager to prioritize events based on their effect on drinking water safety. The special events that are identified in the WSP can be incorporated in QMRA as risk scenarios in order to assess the combined risk of infection of normal and special events.

## Setting critical limits

A treatment system can be designed to provide exactly the right level of treatment to meet the health-based targets. However, in practice the risk manager needs to account for variations and inaccuracies in order to run a practical and stable process. Treatment systems are controlled by setpoints, operational limits and critical limits. During normal operation at the setpoint, the treatment will run between operational limits. When the process deviates beyond the critical limits, corrective actions are required in order to meet health-based targets. Setting of appropriate operational and critical limits is complex since they depend on the (long term) treatment target, variability of the process, response time and the options for corrective actions. The applicable safety margin for treatment efficacy is limited due to other goals such as costs or prevention of disinfection by-products. QMRA can address these issues by quantifying the microbial risk outcome of different options both for individual and combined processes. Arriving at the optimal limits will need several iterations, using practical experience and ongoing scientific insights to further improve the operation of the treatment system. Critical limits will depend on circumstances such as water temperature or source water turbidity. For complicated systems a real-time computer model of the water supply system (for disinfection and other water quality parameters) may be helpful in maintaining optimal water quality and choosing the most appropriate corrective measures.

## Designing monitoring programs

Monitoring of treatment systems serves two goals. On the one hand monitoring is applied to verify that the system nominally meets the health-based targets. Microbial monitoring provides the most direct verification of system performance. However microbial monitoring requires resources and funds, and cannot be applied limitlessly. QMRA can be used to design the microbial monitoring plan so that results will provide a statistically valid verification of treatment performance at the required confidence level. Monitoring results can be used in QMRA to adapt the microbial monitoring program to match the site specific situation.

On the other hand monitoring to detect events requires a high measurement frequency, which is not feasible with microbial monitoring. Rather than quantifying efficacy, this type of monitoring should detect deviations that indicate that treatment is failing. Monitoring of surrogates (turbidity, particles), process conditions (flow, temperature, disinfectant residual) and equipment (dosing pump, valves) can provide an indication of failure and is generally easier and cheaper than microbial monitoring. Very short failure events can significantly impact the mean treatment efficacy. QMRA can be used to design frequency of (on-line) monitoring to verify that the health-based targets are not compromised by failure events.

### Preparing corrective actions

When critical limits are exceeded, corrective actions are needed to restore system control and prevent non-compliance with the health-based target. Different levels of corrective actions may be undertaken. These could be restricted to the control measure that is out of bounds, but could also include other control measures that may be enhanced or additional (emergency) control measures. QMRA can be used to select the most appropriate corrective actions under the given conditions, as it looks at the system as a whole, rather than at individual control measures. The level and duration of the required corrective action can also be determined through QMRA.

### Treatment design: comparing alternatives

During the design of a water treatment plant, or when changes to a treatment plant are required, one needs to choose between different solutions. Each (combination of) solution needs to comply with the health-based targets. A QMRA can help identify the most economic alternative. Thus unnecessary investments can be avoided. Here QMRA can be used as a design tool. QMRA can also be used in the design stage to evaluate control strategies, determine required redundancy and prepare effective monitoring.

## FOCUS OF THIS BOOK

This book deals with the principal question 'How can we quantify the reduction of pathogens by drinking water treatment for QMRA purposes?' by addressing the following questions.

'How can we combine all site specific full-scale information?'
The literature study in this chapter made clear that the many types of data, microbial and non-microbial, could be used to quantify treatment efficacy of a full-scale system. The type of data could vary per treatment process. Therefore the framework to combine different types of data on treatment efficacy in Chapter 1 was developed. Chapter 2 discusses the implementation of the framework in risk assessment.

'What can we learn from microbial monitoring of drinking water'
Although microbial monitoring can only verify drinking water safety to a certain extent, the information collected by microbial monitoring of drinking water does provide a direct impression of microorganism distribution in drinking water. Such datasets can include a large number of non-detects (or 'zeros'). Most studies had used mean concentrations in drinking water derived from these datasets for QMRA, thus disregarding the variability. The study in Chapter 3 provides an example of a stochastic risk assessment based on these data, including variability and assessment of the impact of the non-detects interpretation on the assessed level of risk.

'How can we use process models in QMRA?'
Process models for disinfection processes have been used for legislation. However, in some cases indicator bacteria were detected in disinfected water even when process models predicted indicator concentrations many

orders of magnitude below detection limit. If process models were to be used for QMRA, these need to provide a more accurate estimate of treatment efficacy at full-scale. An ozonation disinfection model that could be applied for full-scale modelling in QMRA was developed in Chapter 4. Chapters 2 and 7 applied disinfection modelling in QMRA.

'How can we include variability and uncertainty in treatment modelling?'
The information on full-scale treatment efficacy was expected to include significant uncertainty, and previous studies had shown that treatment efficacy could vary substantially in time and between sites. The methodology for stochastic modelling of treatment (Teunis *et al.*, 1997, 1999; Teunis and Havelaar, 1999; Medema *et al.*, 1999 and Haas *et al.*, 1999) was developed further in Chapters 5 and 6 in order to include variability and uncertainty from various types of data in treatment modelling.

'How can we quantify treatment efficacy based on microbial monitoring?'
The microbial monitoring data in treated water and at different stages in treatment could provide the best site specific full-scale information to assess treatment efficacy. Available methods (Teunis *et al.*, 1997, 1999; Teunis and Havelaar, 1999; Medema *et al.*, 1999 and Haas *et al.*, 1999) did not always provide an accurate prediction of microorganism concentrations after treatment. An improved methodology for stochastic treatment modelling was developed in Chapter 5.

'How accurate is stochastic treatment modelling in practice?'
Few of the studies that applied stochastic treatment modelling actually validated the model outcomes with monitoring data. None of the studies verified how accurately the stochastic modelling predicted future treatment performance or how well a model calibrated with indicator data predicted pathogen reduction. Chapter 7 assesses the accurateness of stochastic treatment modelling applications for QMRA in practice.

'How can QMRA be applied in the WSP?'
QMRA in itself does not improve drinking water safety. QMRA can be applied as a tool in water safety plans (WSP), as discussed previously in this chapter. QMRA could provide objective, quantitative decision support for risk managers working with the WSP. Chapter 7 provides examples of QMRA applications for use in the WSP.

## APPROACH OF THE PRESENTED STUDY

### Catchment to Tap System (CTS)

The study was started in 2002 as part of the MicroRisk project (MicroRisk, 2002 and Medema *et al.*, 2006) in which 12 partners from 7 countries participated. The goal of the MicroRisk study was to provide a unified approach for QMRA in drinking water from source to tap. The work presented here was part of work package 3 'Treatment'.

Historical treatment data was collected from twelve Catchment to Tap Systems (CTS) during the project. Appropriate datasets were analysed to determine the usability of such data for QMRA. This study especially focussed on the water treatment system of the city of Amsterdam which was studied in detail in two aspects. The first aspect was the development of the QMRA methodology for treatment using the extensive amount of data already available for this treatment system. The second aspect was the optimization of the ozonation process at this location (Chapter 4). Ozonation efficacy was assessed by stochastic process modelling, laboratory experiments, pilot experiments and mathematical process modelling. The treatment assessment framework was developed to effectively combine the data generated in the MicroRisk project.

## The treatment assessment framework

The treatment assessment framework in Figure 1.2 was used to combine information about pathogen reduction by drinking water treatment as input for a QMRA model. Different pathogens pose varying challenges to water treatment. Bacteria are removed less by filtration than other microorganisms but are readily inactivated by disinfection. Protozoa are relatively insensitive to chlorine disinfection but are better removed by filtration than bacteria and readily inactivated by UV. Virus removal and inactivation is in the range between bacteria and protozoa reduction. By assessing the reduction of a suite of pathogens and indicators, the challenges posed by other (unknown) pathogens was covered. The treatment assessment did not differentiate between different strains of microorganisms unless these were specifically discussed.

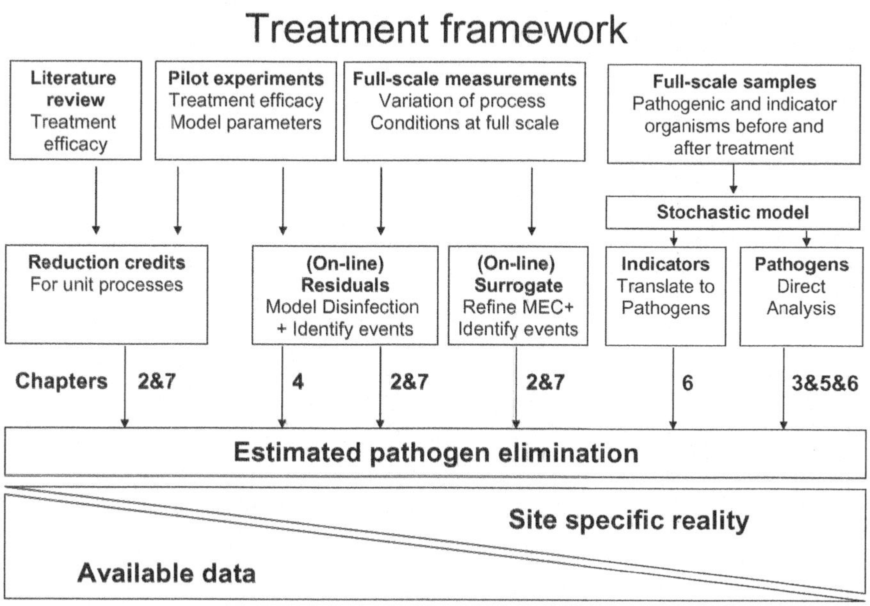


**Figure 1.2** MicroRisk treatment assessment framework. The chapter numbers refer to the chapters where the approach is applied

In order to assess the efficacy of a treatment system, site specific data that provided information on the performance of that system was needed. Frequent monitoring of pathogens at different stages of treatment would provide the ideal dataset to determine the barrier efficiency, but pathogen concentrations are often too low for detection. Especially monitoring at the final stages of treatment would result in mostly non-detects. Various indicator organisms, surrogate parameters or process conditions were used to estimate pathogen treatment efficacy. Many water utilities already collected some of this data either for compliance with drinking water legislation (indicator organism sampling) or for operational purposes (residual chlorine measurement to control chlorine dosing). The available data from the MicroRisk CTSs was compiled to provide an overview of treatment performance for a range of treatment processes under different conditions in different countries. Thus European water treatment practice and data availability was reflected.

Reduction at a treatment site varied in time and this variation was monitored by collecting data over a period of several years. The quality of the data varied between the CTSs, and between treatment processes within a single CTS. Figure 1.2 illustrates how reduction by each treatment barrier can be estimated based on commonly available data. The numbers in Figure 1.2 refer to the chapters in this book that discuss the use of these types of data.

Other types of data that can be used for treatment performance assessment include results from (site specific) pilot tests as discussed for ozonation in Chapter 4. Such tests are more applicable to the local situation than general literature values. Use of other data like operational diaries or failure reports can provide information about the frequency and duration of events in treatment (Westrell *et al.*, 2003; Nilsson *et al.*, 2007 and Signor *et al.*, 2007).

## REFERENCES


Anonymous (2001). Besluit van 9 januari 2001 tot wijziging van het waterleidingbesluit in verband met de richtlijn betreffende de kwaliteit van voor menselijke consumptie bestemd water. (Adaptation of Dutch drinking water legislation) *Staatsblad van het Koninkrijk der Nederlanden*, **31**, 1–53.

AWWA (2006). http://www.awwa.org/Advocacy/learn/info/HistoryofDrinkingWater.cfm (accessed 1 March 2008).

Barry S. J., Atwill E. R., Tate K. W., Koopman T. S., Cullor J. and Huff T. (1998). Developing and Implementing a HACCP-Based Programme to Control *Cryptosporidium* and Other Waterborne Pathogens in the Alameda Creek Watershed: Case Study. American Water Works Association Annual Conference, 21–25 June 1998, Dallas, Texas, *Water Resources*, **B**, 57–69.

Brock T. D. (1988). *Robert Koch: a life in medicine and bacteriology*. Science Tech. Publishers, New York, USA.

Bukowski J., Korn L. and Wartenberg D. (1995). Correlated inputs in quantitative risk assessment: the effects of distributional shape. *Risk Analysis*, **15**(2), 215–219.

Crabtree K. D., Gerba C. P., Rose J. B. and Haas C. N. (1997). Waterborne Adenovirus: A risk assessment. *Water Science and Technology*, **35**(11–12), 1–6.

CDW (2007). *Approach for CDW Discussion and Decision; Microbiological Documents – Enteric Viruses and Protozoa* Draft Discussion Document – ITEM 3.04 October 2007. Federal-Provincial-Territorial Committee on Drinking Water (CDW) Health Canada, Ottawa (Ontario), Canada.

Davison A., Deere D., Stevens M., Howard G. and Bartram J. (2006). *Water Safety Plan Manual*. World health organization, Geneva, Switzerland.

Dewettinck T., Van Houtte E., Geenens D., Van Hege K. and Verstraete W. (2001). HACCP (Hazard Analysis and Critical Control Points) to guarantee safe water reuse and drinking water production – a case study. *Water Science and Technology*, **43**(12), 31–38.

Deere D. and Davison A. (1998). Safe water – are food guidelines the answer? *Water*, **25**, 21–24.

DWI, Drinking Water Inspectorate (1999). *UK Water Supply Regulations (Amended)*, www.dwi.gov.uk (accessed 1 March 2008).

Fewtrell L. MacGill S. M., Kay D. and Casemore D. (2001). Uncertainties in risk assessment for the determination of drinking water pollutant concentrations: *Cryptosporidium* case study. *Water Research*, **35**(2), 441–447.

Gale P., Van Dijk P. A. H. and Stanfield G. (1997). Drinking water treatment increases microorganism clustering: The implications for microbiological risk assessment. *Journal of Water Supply Research and Technology – Aqua*, **46**, 117–126.

Gale P. (2001). Developments in microbiological risk assessment for drinking water. *Journal of Applied Microbiology*, **91**, 191–205.

Gibson III C. J., Haas C. N. and Rose J. B. (1999). Risk assessment of waterborne protozoa: current status and future trends. *Parasitology*, **117**, 205–212.

Gerba C. P. and Haas C. N. (1988). Assessment of risks associated with enteric viruses in contaminated drinking water. *ASTM Special Technical Publication*, **976**, 489–494.

Gerba C. P., Rose J. B., Haas C. N. and Crabtree K. D. (1996). Waterborne rotavirus: a risk assessment. *Water Research*, **30**(12), 2929–2940.

Greenwood M. and Yule G. U. (1917). On the statistical interpretation of some bacteriological methods employed in water analysis. *Journal of Hygiene*, **16**, 36–54.

Haas C. N. (1983). Estimation of risk due to low doses of microorganisms: A comparison of alternative methodologies. *American Journal of Epidemiology*, **118**, 573–582.

Haas C. N., Rose J. B., Gerba C. P. and Regli S. (1993). Risk assessment of virus in drinking water. *Risk Analysis*, **13**(5), 545–552.

Haas C. N., Rose J. B. and Gerba C. P. (1999). *Quantitative Microbial Risk Assessment.* Wiley, New York, USA.

Haas C. N. (1999). On modeling correlated random variables in risk assessment. *Risk Analysis*, **19**(6), 1205–1214.

Haas C. N. and Trussel R. R. (1998). Frameworks for assessing reliability of multiple, independent barriers in potable water reuse. *Water Science and Technology*, **38**(6), 1–8.

Haas C. N. and Kaymak B. (2003). Effect of initial microbial density on inactivation of *Giardia muris* by ozone. *Water Research*, **37**, 2980–2988.

Havelaar A. H. (1994). Application of HACCP to drinking water supply. *Food Control*, **5**(3), 145–152.

Hellard M. E., Sinclair M. I., Forbes A. B. and Fairley C. K. (2001). A randomized, blinded, controlled trial investigating the gastrointestinal health effects of drinking water quality. *Environmental Health Perspectives*, **109**(8), 773–778.

Hrudey S. E., and Hrudey E. J. (2004). *Safe Drinking Water – Lessons Learned from Recent Outbreaks in Affluent Nations.* IWA Publishing, London.

ILSI Risk Science Institute Pathogen Risk Assessment Working Group (1996). A conceptual framework to assess the risks of human disease following exposure to pathogens. *Risk Analysis*, **16**(6), 841–848.

Isaac-Renton J., Blatherwick J., Bowie W. R., Fyfe M., Khan M., Li A., King A., McLean M., Medd L., Moorehead W., Ong C. and Robertson W. (1999). Epidemic and endemic seroprevalence of antibodies to *Cryptosporidium* and *Giardia* in residents of three communities with different drinking water supplies. *American Journal of Tropical Medicine and Hygiene*, **60**(4), 578–583.

IWA/WHO (2004). *IWA Bonn Charter for Safe Drinking Water*, IWA publishing, London.

LeChevallier M. W., Norton W. D., Lee R. G., (1991). *Giardia* and *Cryptosporidium* spp. in filtered drinking water supplies. *Applied Environmental Microbiology*, **57**(9), 2617–2621.

LeChevallier M. and Au K. K. (2004). *Water Treatment and Pathogen Control Process: Efficiency in Achieving Safe Drinking Water.* WHO, IWA publishing, London, UK.

Lloyd A. and Drury D. (2002). Continuous monitoring for *Cryptosporidium* – a novel approach to public health protection. *Water Science & Technology*, **46**(11–12), 297–301.

Masago Y., Katayama H., Hashimoto A., Hirata T. and Ohgaki S. (2002). Assessment of risk of infection due to *Cryptosporidium parvum* in drinking water. *Water Science and Technology*, **46**(11–12), 319–324.

Medema G. J. Groennou J., Hijnen W. A. M. Teunis P. Kruidenier L., Ketelaars H., Willemsen-Zwaagstra J, Havelaar A. and Van der Kooij D. (1999). Frequency distributions of *Cryptosporidium* and *Giardia* in raw water and elimination capacity of water treatment. In: Cryptosporidium *and* Giardia: *new challenges to the water industry*, Medema G. J. (ed.). Thesis, University of Utrecht, the Netherlands.

Medema G. J., Loret J. F., Stenström T. A. and Ashbolt N. (eds.). (2006). *Quantitative Microbial Risk Assessment in the Water Safety Plan*, report for the European Commission under the Fifth Framework Programme, Theme 4: 'Energy, environment and sustainable development' (contract EVK1CT200200123), Kiwa Water Research, Nieuwegein, the Netherlands.

MicroRisk (2002). www.microrisk.com (accessed 1 March 2008).

Nichols G. (2003). Using Existing Surveillance-Based Data. In: *Drinking Water and Infectious Disease: Establishing the Links*, Hunter P. R., Waite M., Ronchi E. IWA Publishing, London, UK.

Nilsson P., Roser D., Thorwaldsdotter R., Petterson S., Davies C., Signor R., Bergstedt O. and Ashbolt N. (2007). SCADA data and the quantification of hazardous events for QMRA. *Journal of Water and Health*, **5**(S1), 99–105.

NWQM (2006, 2007) *Australian Guidelines for Water Recycling. Managing Health and Environmental Risks (Phase 1).* www.ephc.gov.au/ephc/water_recycling.html (accessed 1 March 2008).

Payment P., Trudel M. and Plante R. (1985). Elimination of viruses and indicator bacteria at each step of treatment during preparation of drinking water at seven water treatment plants. *Applied Environmental Microbiology*, **49**, 1418–1428.

Payment P. and Franco E. (1993). *Clostridium perfringens* and somatic coliphages as indicators of the efficiency of drinking water treatment for viruses and protozoan cysts. *Applied and Environmental Microbiology*, **59**(8), 2418–2424.

Payment P., Berte A., Prévost M., Ménard B. and Barbeau B. (2000). Occurrence of pathogenic microorganisms in the Saint Lawrence River (Canada) and comparison of health risks for populations using it as their source of drinking water. *Canadian Journal of Microbiology*, **46**, 565–576.

Pipes W. O., Ward P. and Ahn S. H. (1977). Frequency distributions for coliform bacteria in water. *Journal of American Waterworks Association JAWWA*, **69**, 664–668.

Regli S., Rose J. B., Haas C. N. and Gerba C. P. (1991). Modeling the Risk from *Giardia* and viruses in drinking water. *Journal of the American Water Works Association JAWWA*, **83**(11), 76–84.

Rose J. B., Haas C. N. and Regli S. (1991). Risk assessment and the control of waterborne *Giardiasis*. *American Journal of Public Health*, **81**(6), 709–713.

Rose J. B. and Masago Y. (2007). A toast to our health: our journey toward safe water. *Water Science & Technology: Water Supply*, **7**(1), 41–48.

SDWA (1974). *Safe Drinking Water Act of 1974*. Public Law 93-523, Dec. 16, 1974. Washington, DC: US Government Printing Office, SDWA, 1977.

Signor R. S., Ashbolt N. J. and Roser D. J. (2007). Microbial risk implications of rainfall-induced runoff events entering a reservoir used as a drinking-water source. *Journal of Water Supply – AQUA*, **56**, 515–531.

Signor R. S. and Ashbolt N. J. (2007). On comparing probabilistic microbial risk assessments against daily rather than annualized infection probability health-targets, *14th Health-Related Water Microbiology Symposium*. International Water Association, Tokyo, Japan.

Smith A. E., Ryan P. B. and Evans J. S. (1992). The effect of neglecting correlations when propagating uncertainty and estimating the population distribution of risk. *Risk Analysis*, **12**(4), 467–474.

Taubes G. (1995). Epidemiology faces its limits. *Science*, **269**(5221), 164–169.

Teunis P. F. M., Havelaar A. H. and Medema G. J. (1994). *A literature Survey on the Assessment of Microbiological Risk for Drinking Water*. Report 734301006, RIVM Bilthoven, the Netherlands.

Teunis P. F. M., Medema G. J., Kruidenier L. and Havelaar A. H. (1997). Assessment of the risk of infection by *Cryptosporidium* or *Giardia* in drinking water from a surface water source. *Water Research*, **31**(6), 1333–1346.

Teunis P. F. M. and Havelaar A. H. (1999). *Cryptosporidium* in drinking water: Evaluation of the ILSI/RSI quantitative risk assessment framework. Report No. 284550006 RIVM, Bilthoven, the Netherlands.

Teunis P. F. M, Evers E. G. and Slob W. (1999). Analysis of variable fractions resulting from microbial counts. *Quantitative Microbiology*, **1**, 63–88.

Teunis P., Davison A. and Deere D. (2004). *Short-term Fluctuations in Drinking Water Quality and their Significance for Public Health*. World health organization, Geneva, Switzerland.

USEPA (1989). *SWTR, Surface Water Treatment Rule*. United States Environmental Protection Agency, Washington, DC, USA.

USEPA (1998). *IESWTR, Interim Enhanced Surface Water Treatment Rule*. United States Environmental Protection Agency, Washington, DC, USA.

USEPA (2001). *Drinking water Criteria Document addendum: Cryptosporidium*. United States Environmental Protection Agency, Washington, DC, USA.

USEPA (2002). *LT1ESWTR, Long Term First Enhanced Surface Water Treatment Rule*. United States Environmental Protection Agency, Washington, DC, USA.

USEPA (2006). *LT2ESWTR, Long Term Second Enhanced Surface Water Treatment Rule*. United States Environmental Protection Agency, Washington, DC, USA.

USEPA (2007). *Report of the experts scientific workshop on critical research needs for the development of new or revised recreational water quality criteria*. Airlie Center Warrenton, Virginia March 26–30, 2007 EPA 823-R-07-006 United States Environmental Protection Agency, Washington, DC, USA.

Westrell T., Bergstedt O., Stenström T. A. and Ashbolt N. J. (2003). A theoretical approach to assess microbial risks due to failures in drinking water systems International. *Journal of Environmental Health Research*, **13**(2), 181–197.

WHO (2001). *Water Quality: Guidelines, Standards and Health*. World health organization, Geneva, Switzerland.

WHO (2004). *Guidelines for Drinking Water Quality (third edition)*. World health organization, Geneva, Switzerland.

# Chapter 2

# A stochastic pathogen reduction model for full-scale treatment

P.W.M.H. Smeets[a,b], L.C. Rietveld[b], J.C. van Dijk[b] and G.J. Medema[a,b]

[a]KWR Watercycle research institute, PO Box 1072, 3430 BB Nieuwegein, the Netherlands
[b]Faculty of Civil Engineering, Delft University of Technology, PO Box 5048, 2600 GA Delft, the Netherlands

## ABSTRACT

Determining the concentration of pathogens in drinking water is a crucial step in quantitative microbial risk assessment of drinking water. When pathogens cannot be detected directly in drinking water, their concentration in drinking water can be calculated from their concentration in raw water and their reduction by treatment. This study compared a point estimate approach to stochastic approaches for this calculation using different types of data in a case study of *Campylobacter* removal by surface water treatment with rapid sand filtration, ozonation and slow sand filtration. The point estimate based on literature, provided an indication of the mean *Campylobacter* concentration in drinking water. The point estimate also showed that the *Campylobacter* concentration could be several orders of magnitude higher or lower than the average under extreme conditions. How certain the estimate of the mean concentration was, or how often extreme conditions occurred was not assessed. Therefore three stochastic assessments were performed.

A stochastic assessment using full-scale *Campylobacter* monitoring data incorporated the variability of raw water concentration and treatment efficacy. This provided an estimate of frequency and magnitude of peak *Campylobacter* concentrations in drinking water. In addition the stochastic method provided insight

 *Stochastic Modelling of Drinking Water Treatment in Quantitative Microbial Risk Assessment.* By Patrick Smeets. ISBN: 9781843393740. Published by IWA Publishing, London, UK.

in the uncertainty of the estimated raw water concentration and treatment efficacy. As an alternative, *E. coli* monitoring data was used as a surrogate for *Campylobacter*. This more extensive data provided a more detailed insight in the estimated variation. Finally the inactivation by ozonation was calculated with a process model using monitored ozone concentrations, temperatures and contact times to illustrate a situation where microbial monitoring was not feasible. The process model results were similar to the microbial assessments. The stochastic modelling in this study included the variability of the system. However, the assessed uncertainty about raw water concentrations and treatment efficacies could affect the calculated *Campylobacter* concentrations in drinking water. In addition, risk managers and legislators need to know the uncertainty of the assessment when using it as a basis for decisions. Future research should therefore focus on including uncertainty in the stochastic assessment.

## INTRODUCTION

Water treatment is the major barrier against pathogenic microorganisms when preparing drinking water from surface water. The level of treatment is to be adapted to the pathogen contamination level of the source water. An estimate of the number of pathogens in the treated water can be calculated from the measured amount of pathogens in source water and the estimated treatment efficacy as part of a risk assessment (Regli *et al.*, 1991 and Haas *et al.*, 1999). In a point estimate assessment the efficacy of a treatment step is represented by a single value for efficacy (USEPA, 2003 and WHO, 2004). However, it is known that full-scale treatment efficacy varies in time due to variation of conditions such as water quality, flows and operation. Short term peaks of pathogens in the source water or a short period of failure in treatment can lead to an increase of the average exposure and thus to an increase of the risk of infection (Teunis *et al.*, 2005). Stochastic methods have been proposed to calculate pathogen concentration in the treated water using varying raw water concentrations and treatment efficacy (Bartram *et al.*, 2001 and Medema, 2003). In a stochastic assessment, the treatment efficacy is represented by a variable value. The first goal of the study was to compare the point estimate assessment to the stochastic assessment.

Stochastic assessments have used monitoring data of pathogens or indicators in raw water and water at different stages of treatment (Teunis *et al.*, 1997).

A large dataset of pathogen analysis before and after treatment would be ideal for treatment assessment. However, pathogen analyses are generally scarce and after a certain point in treatment pathogens can no longer be detected. Since indicator organisms such as *E. coli* are present in the source water in higher concentrations than pathogens, they can often be detected further down the treatment train. This, and the fact that they are more frequently measured for legislative purposes makes the indicator data valuable for treatment assessment. The second goal of the study was to compare an assessment with pathogen data to an assessment with indicator organism data.

In cases where the removal of pathogens can be modelled with a process model using input from process parameters, information about process parameters can provide estimates of treatment efficacy. In addition, process conditions such as disinfectant residual, temperature and flow can be monitored on-line. This provides an opportunity for detailed assessment of treatment variability at relatively low costs. The third goal of the study was to determine whether a process model could provide a realistic estimate of treatment efficacy in an assessment.

## MATERIALS AND METHODS

### Case study system

The case study used data from a full-scale treatment system. River water was pre-treated by coagulation, sedimentation and filtration. The pre-treated water was transported to the dunes for artificial recharge in open canals. After a residence time of approximately 90 days, the water was abstracted through open canals and stored in an open pond. Since the soil passage removed most microorganisms from the pre-treated water, recontamination in the pond by waterfowl and wildlife was the main source of faecal contamination. Since the water in the pond was the last open water before final treatment, it was considered as the raw water in the treatment assessment. The final treatment system is shown in Figure 2.1. The water was abstracted from the pond (RAW) and treated by rapid sand filtration (RSF), ozonation (O3) and slow sand filtration (SSF).

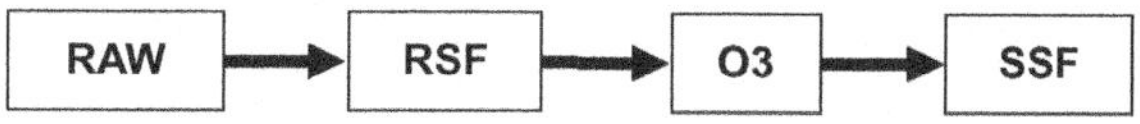


**Figure 2.1** Treatment steps of the case study treatment system

Available data consisted of *Campylobacter* and *E. coli* measurements in raw, filtered and ozonated water over an eight-year period and measurements of ozone process conditions. The system consisted of five parallel lines of six ozone contact chambers in series. Temperature was measured in the main flow at the intake point. Flow was measured for each of the five lines. Ozone concentration was measured in each of the thirty contact chambers.

## Microbial monitoring

*Campylobacter* was analysed according to NEN 6269 (Anonymous, 1996). Samples were analysed by direct filtration and direct inoculation of the filter in tubes with Preston *Campylobacter* selective enrichment broth. The tubes were incubated in microaerobic conditions for 48 hr at 42°C. A loopful of Preston medium was then transferred onto Karmali selective medium. These pallets were incubated for 48 hr at 42°C. Positive results were confirmed by microscopic examination in a hanging drop for the typical morphology and motility of *Campylobacter*. *Campylobacter* was quantified by the most probable number method (MPN) in three parallel tubes for three filtered sample volumes using decimal dilutions. Reported MPN concentrations were taken from MPN tables by De Man (1975). *E. coli* was analysed by direct filtration and direct inoculation of the filter on lauryl sulphate agar (Oxoid Basingstoke, England, nr. MM0615). The plates were incubated for 4 hr at 25°C and then for 18–24 hr at 44°C. A proportion of the typical yellow colonies were confirmed in brilliant green bile salts lactose medium for 24–48 hr at 44°C.

## Pathogen reduction model

Pathogen concentrations in treated water $C_{out}$ were calculated from the pathogen concentration in raw water $C_{raw}$ and their reduction by the treatment processes in series using Equation 1 (Haas *et al.*, 1999).

$$C_{out} = C_{raw} \times \pi_{rsf} \times \pi_{O3} \times \pi_{ssf} \qquad (1)$$

Where $\pi_{rsf}$ is the efficacy of rapid sand filtration, $\pi_{O3}$ is the efficacy of ozonation and $\pi_{ssf}$ is the efficacy of slow sand filtration. Treatment efficacy $\pi$ was the fraction of organisms that passed a treatment step. The (variation of) removal efficacy $\pi$ was taken from literature, calculated with a process model or calculated from monitored concentrations of pathogens or indicator organisms before and after a treatment process on the same day using Equation 2.

$$\pi_d = \frac{C_{d_{out}}}{C_{d_{in}}} \qquad (2)$$

where $C_{d_{out}}$ is the concentration of organisms after a treatment step on date *d* and $C_{d_{in}}$ is the concentration of organisms before a treatment step on date *d*.

Treatment efficacy was expressed as log reduction $\log_{10}(\pi)$. The log reduction over a longer period was calculated from the arithmetic mean concentrations before and after treatment over the total period with Equation 3.

$$DE = -\log_{10}\left(\frac{\overline{C_{out}}}{\overline{C_{in}}}\right) \tag{3}$$

## Process model for ozonation

Disinfection by ozonation was modelled using ozone concentration $C_{O3}$, contact time *t* and inactivation rate constant $k_e$. The CT10 model (Equation 4) and the CSTR model (Equation 5) were used (Smeets *et al.*, 2006).

$$\frac{C_{out}}{C_{in}} = e^{-k_e * C_{O3} t_{10}} \tag{4}$$

$$\frac{C_{out}}{C_{in}} = \frac{1}{1 + k_e C_{O3} t} \tag{5}$$

In which $C_{O3}$ was the ozone concentration, $t_{10}$ was the contact time that was exceeded by 90% of the water. Contact time *t* was calculated from the measured flow and the volume of the contactor and $t_{10}/t$ was estimated as 0.56 based on tracer tests. The CSTR disinfection model for this system was discussed by Smeets *et al.* (2006). Due to ozone decay, ozone was not detected in all chambers all the time. The appropriate number of CSTR in series therefore varied between 1 and 6. The point estimate in approach 1 used the mean number of 3 CSTR as a best estimate and 1 and 6 CSTR as extreme values. In approach 4 each of the thirty contact chambers was modelled as a CSTR with Equation 5 using the measured ozone residual, flow and temperature. Contact time *t* was calculated from the flow per treatment line and the volume of each contact chamber. The efficacy of each line was calculated by multiplying the efficacy of each contactor in that line. The efficacy of the total ozonation process was calculated as the mean of the individual lines on each date. A beta PDF was fitted to the calculated inactivations.

## Point estimate assessment

Discrete values were used in Equation 1 in the point estimate assessment. The point estimate of raw water *Campylobacter* concentration was calculated as the arithmetic mean of the monitored concentrations at the intake. The values for $\pi_{rsf}$ and $\pi_{ssf}$ were taken from a literature review including pilot experiments and full-scale data analysis (Hijnen *et al.*, 2005a, 2005b). $\pi_{O3}$ was calculated with Equations 4 and 5 using the average operational setpoints. Equation 1 was calculated with the mean, maximum and minimum reported values for each step in the assessment. This resulted in a best estimate, a maximum estimate and a minimum estimate of the *Campylobacter* concentration in treated water. A sensitivity analysis of the point estimate was performed by making all possible combinations of the mean, maximum and minimum values of the assessment steps.

## Stochastic assessment

The variability of $C_{in}$, $\pi_{rsf}$, $\pi_{O3}$ and $\pi_{ssf}$ were addressed in a stochastic assessment by Monte Carlo simulation of Equation 1. Each variable was represented by a probability density function (PDF). The raw water concentrations were described by fitting a gamma PDF to the monitored *Campylobacter* concentrations (Haas *et al.*, 1999). A beta PDF was fitted to the removal values of $\pi_{rsf}$ and $\pi_{O3}$ assessed from the monitoring data (Teunis *et al.*, 1999). The maximum likelihood estimates (MLE) of the fitted PDF parameters were determined by the 'pdf-fit' functions in Matlab® statistical toolbox. This function also provided the parameter pairs $\alpha_{2.5\%} - \beta_{2.5\%}$ and $\alpha_{97.5\%} - \beta_{97.5\%}$ of the 95% confidence interval (CI) boundaries of the fitted PDF. The 95% CI parameters thus indicated how well the PDF described the data. No site specific data was available for the slow sand filtration. Therefore the values of $\pi_{ssf}$ reported in literature (Hijnen *et al.*, 2005a, 2005b) were used as if they represented the variability of efficacy (Figure 2.2). A beta PDF was fitted to the $\pi_{ssf}$ data and used in the Monte Carlo simulation. Due to the limited amount of data points, the 95% CI boundaries of the PDF parameters could not be determined.

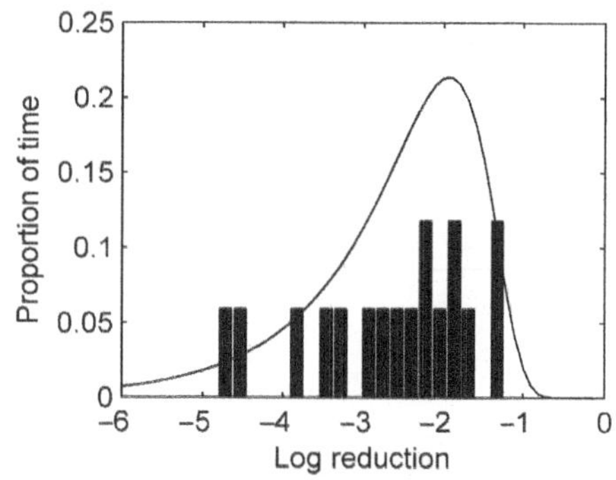


**Figure 2.2** Removal of bacteria by slow sand filtration as reported in literature (Hijnen *et al.*, 2005a, 2005b) (bars) and the fitted beta PDF (line)

Monte Carlo simulation was performed by drawing random values from each PDF of $C_{in}$, $\pi_{rsf}$, $\pi_{O3}$ and $\pi_{ssf}$. The pathogen concentration after treatment was calculated from the random values with Equation 1. This procedure was repeated 100,000 times, resulting in as many estimates of *Campylobacter* concentration in drinking water. Together these estimates described the variability of the estimated concentration.

## Approaches

Four different approaches were applied to study the impact of choices made in the risk assessment. Approach 1 used the point estimate of the raw water concentration and the efficacy of each treatment step. Ozonation was efficacy was estimated with the CSTR model (Equation 5) based on operational settings. Approach 1 reflected the situation when no site specific data on treatment efficacy was available and a 'simple' calculation method was chosen. Approach 2 used Monte Carlo simulation with the PDFs based on the *Campylobacter* (pathogen) data for $C_{in}$, $\pi_{rsf}$, and $\pi_{O3}$ and the PDF based on literature for $\pi_{ssf}$. This reflected a situation where some microbial monitoring data was available and variability was taken into account. Approach 3 used the large number of *E. coli* data as an indicator of variation of $C_{in}$ and as a surrogate for *Campylobacter* reduction $\pi_{rsf}$, and $\pi_{O3}$. The PDF based on literature was used for $\pi_{ssf}$. This assessment showed the effect of using a large dataset in the treatment assessment. Approach 4 was similar to approach 3 for $C_{in}$, $\pi_{rsf}$, and $\pi_{ssf}$. Ozone efficacy

$\pi_{O3}$ was modelled with the CSTR model (Equation 5) using monitored ozone concentrations, temperatures and contact times (based on flow). This illustrated how a disinfection process could be included in a treatment assessment when microbial monitoring was not feasible.

# RESULTS

## Microbial monitoring results

The study used long-term water quality monitoring data of *Campylobacter*, *E. coli*, temperature, ozone contact time and residual ozone concentration from the full-scale plant over a six year period. Table 2.1 summarizes the data.

**Table 2.1** Overview of monitoring data

| | *Campylobacter* | | | *E. coli* | | | Ozonation conditions | | | |
|---|---|---|---|---|---|---|---|---|---|---|
| | # Data | Mean MPN/L | % positive | # Data | Mean CFU/L | % positive | | # Data | Mean | Min–Max |
| Raw | 33 | 199 | 100% | 1963 | 383 | 99% | $T$ (°C) | 354 | 12 | 2.3–19.5 |
| Filtered | 31 | 12 | 100% | 1988 | 29 | 84% | $C_{O3}$ (mg/L) | 354 | 0.2 | 0–0.8 |
| Ozonated | 31 | 0.2 | 35% | 3447 | 0.18 | 5% | $t$ (min) | 354 | 10 | 7–13 |

## Approach 1: Point estimate

The arithmetic mean *Campylobacter* concentration in raw water was 199 MPN/L. The minimum and maximum monitored concentrations of 0.3 and 1100 *Campylobacter*/L were used in the point estimate assessment as 'best case' and 'worst case' values. Efficacy of rapid sand filtration for the removal of bacteria was modelled as a point estimate of 0.6 log reduction, based on a literature study (Hijnen *et al.*, 2005b). The minimum and maximum efficacies reported in literature were 0.1 and 1.5 log reduction. The efficacy of ozonation depends on operational conditions. Efficacy of ozonation was estimated from the setpoint of ozonation of a $Ct_{10}$ of 1.2 mg.min/l at a temperature $T$ of 5°C and an inactivation rate constant $k_e$ of 120 L/mg.min (Smeets *et al.*, 2005). The CT10 model predicted over 100 log inactivation of *E. coli*. However, it is known that these high levels of inactivation are not achieved in a conventional ozone contactor due to hydraulic shortcomings (Do-Quang *et al.*, 2000 and Smeets *et al.*, 2006). The more conservative CSTR model approach assuming 6 CSTRs in series and equal spread of CT over the contactors resulted in 8 log inactivation which was used as the maximum estimate of treatment efficacy. The best estimate of 3 CSTR in series resulted in 5 log inactivation and a worst case estimate of 1 CSTR resulted in 2 log inactivation. Efficacy of slow sand filtration was modelled as a point estimate of 2.7 log reduction with a minimum and maximum efficacy of 1.2 and 4.8 log reduction based on a literature study (Hijnen *et al.*, 2005a).

Table 2.2 shows the result of the best estimate and the extreme combinations (minimum and maximum) of treatment step efficacies. This shows that the estimated *Campylobacter* concentrations after treatment ranged over 16 log.

Since there were 4 variables with each 3 estimated values, 81 different combinations were made in the sensitivity analysis of the point estimate assessment. Figure 2.3 shows the calculated *Campylobacter* concentrations in treated water of all these combinations in a histogram. 38% of the combinations led to *Campylobacter* concentrations exceeding the best estimate of $10^{-6}$ *Campylobacter*/L, indicating that the best estimate could underestimate the risk.

**Table 2.2** Results of point-estimate assessment in approach 1

| | min estimate | best estimate | max estimate |
|---|---|---|---|
| source | 0.3 | 199 | 1100 |
| filtration | 1.5 | 0.6 | 0.1 |
| ozonation | 10 | 5 | 2 |
| slow sand filtration | 4.8 | 2.7 | 1.2 |
| total reduction | 16.3 | 8.3 | 3.3 |
| log pathogen concentration | −16.8 | −6.0 | −0.3 |

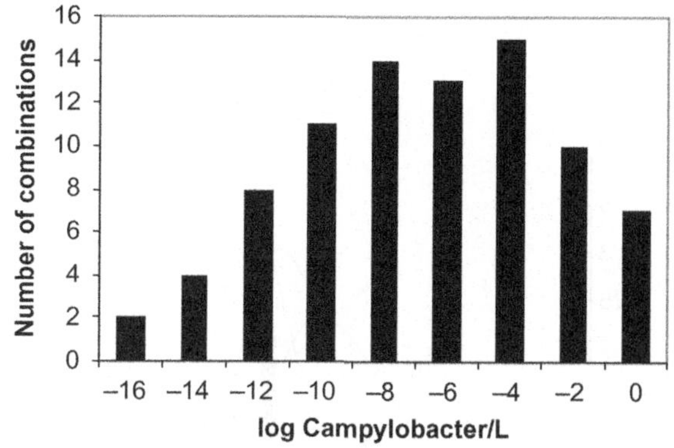


**Figure 2.3** Histogram of calculated mean *Campylobacter* concentration in treated water using combinations of minimum, mean and maximum estimates in the sensitivity analysis of the point estimate assessment

## Approach 2: Pathogen data

Monitored *Campylobacter* concentrations in raw, filtered and ozonated water are shown in Figure 2.4. The reported concentrations ranged over two orders of magnitude, indicating large variations of faecal contamination of the raw water.

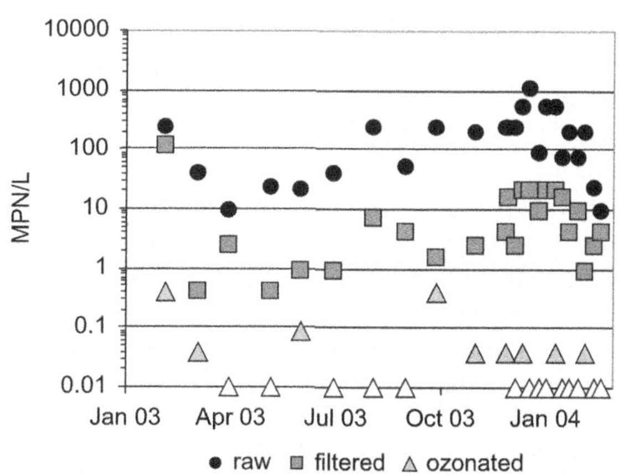


**Figure 2.4** Reported *Campylobacter* concentrations in raw, filtered and ozonated water. Negative samples are shown as open markers at 0.01 MPN/L

A gamma PDF was fitted to the *Campylobacter* concentrations in source water as shown in Figure 2.5. The MLE PDF described the most likely variability of the concentration given the monitoring results. According to this PDF the concentration varied between 1.5 and 853 MPN/L for 95% of the time. The concentration that was exceeded for one day per year (the 99.7th percentile) was 1,434 MPN/l. As a consequence of this variation, the arithmetic mean concentration was 199 MPN/L.

The 95% boundary values of the PDF parameters indicate the uncertainty of how well the PDF described the data. For example, given the monitoring results, the probability that the concentrations in raw water exceeded the concentrations described by the 97.5% PDF was 2.5%. The 97.5% PDF had a mean concentration of 535 MPN/L. This means that there is a 2.5% probability that the mean concentration exceeded 535 MPN/L. Similarly there was a 2.5% probability that the mean concentration was below 74 MPN/L. This uncertainty about the accurateness of the PDF was summarized in Table 2.3 by the mean concentration of the 2.5% and 97.5% PDF.

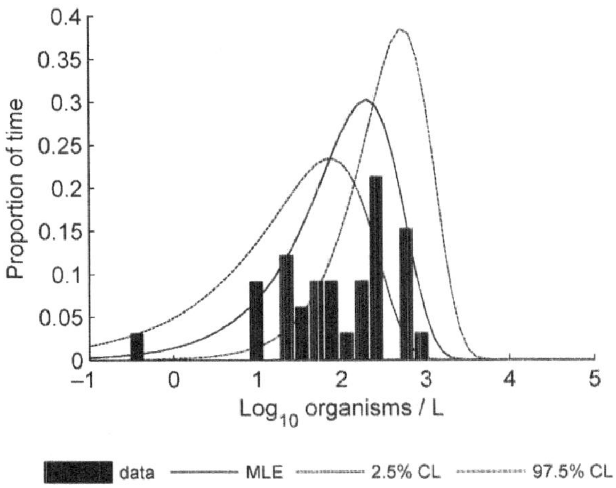


**Figure 2.5** Raw water *Campylobacter* concentrations (histogram) and the gamma probability density function (PDF) with maximum likely parameters (line) and 95% confidence interval of parameters (dashed and dash-dot)

Efficacy of filtration and ozonation were assessed from the monitored *Campylobacter* concentrations before and after these processes. No *Campylobacter* data for slow sand filtration were available, therefore the PDF based on literature values in Figure 2.2 was used in the assessment.

Table 2.3 shows the results of the assessment based on the monitored *Campylobacter* concentrations. The estimated mean raw water concentration was 199 MPN/L, identical to approach 1. According to the fitted PDF, the raw water concentration varied between 1.5 and 853 MPN/L for 95% of the time, which is less extreme than the minimum and maximum estimates in approach 1. Based on the *Campylobacter* data, removal by filtration at this site was more effective (1.1 versus 0.6 log removal) than was expected based on literature data (Table 2.2). The assessed inactivation by ozone was much lower than in approach 1 (1.6 versus 5 log inactivation). Apparently the assumed conditions for the process model in the point estimate did not reflect the actual conditions since some *Campylobacter* were detected after ozonation.

The stochastic assessment provided an estimate of the uncertainties. Table 2.3 shows that the uncertainty with respect to the average raw water concentration ranged from 74 to 535 MPN/L. The uncertainty about the mean efficacy of filtration and ozonation was small (1.0 to 1.3 and 1.5 to 1.8 respectively).

**Table 2.3** Results of pathogen assessment. $\alpha$ and $\beta$ are the MLE parameters of the fitted PDF

| | lower 95% CI boundary | MLE | upper 95% CI boundary | $\alpha$ MLE | $\beta$ MLE | PDF type |
|---|---|---|---|---|---|---|
| raw water (95% CI) MPN/L | 74 | 199 | 535 | 0.72 | 278 | gamma |
| filtration log reduction | 1.3 | 1.1 | 1.0 | 0.70 | 7.65 | beta |
| ozonation log reduction | 1.8 | 1.6 | 1.5 | 0.53 | 19.5 | beta |
| slow sand filtration log reduction | * | 2.0 | * | 0.41 | 36 | beta |

**No parameters for the 95% CI boundaries of the PDF could be determined.*

Figure 2.6 and Table 2.4 show the result of the Monte Carlo simulation. The histogram shows the estimated variability of the *Campylobacter* concentration in treated water. The arithmetic mean *Campylobacter* concentration, due to this variability, was 0.0025 MPN/L (−2.4 log MPN/L). This concentration was 3.6 log higher than the best point-estimate. The difference was caused by the difference in assessed ozonation efficacy.

The stochastic assessment provided an estimate of how the *Campylobacter* concentration in drinking water could vary. The median concentration was 0.000055 MPN/L (−4.3 log MPN/L). The large difference between the median and the mean concentrations was caused by the skewed distribution of concentration. Although high concentrations rarely occurred, they had a high impact on the mean concentration. The mean concentration was exceeded only 12% of the time (88th percentile). The 95th percentile, often used as a measure of peak concentration, was estimated to be 0.0096 MPN/L (−2 log MPN/L). The 99.7th percentile (1−(1/365)) corresponds to the 'maximum day', which was calculated as 0.24 MPN/L. This concentration was expected to occur 0.3% of the time, however this could be spread over several shorter episodes over the year. The maximum estimate of 0.5 MPN/L in approach 1 was of the same order of magnitude. The minimum point estimate of $10^{-16.8}$ did not correspond to the 0.3rd percentile ('minimum day') of $10^{-11}$ of the stochastic assessment.

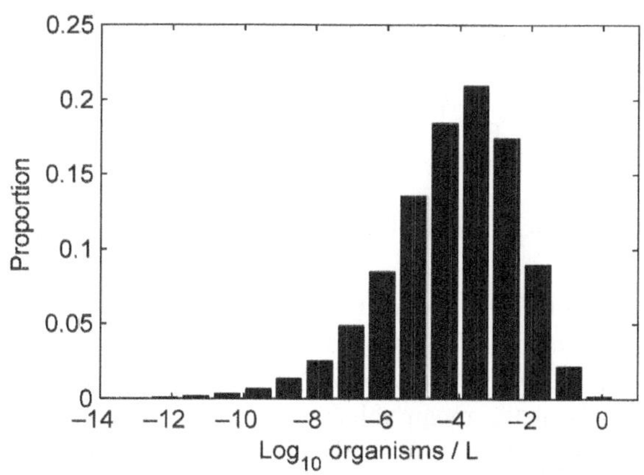


**Figure 2.6** Histogram of log *Campylobacter* concentrations in treated water calculated with Monte Carlo simulation in Approach 2

## Approach 3: Indicator organism data

Figure 2.7 shows the yearly variation of *Campylobacter* and *E. coli* concentration in raw water. The 34 *Campylobacter* concentrations in Figure 2.7 showed a yearly pattern similar to the 1,963 *E. coli* concentrations.

However, the mean *Campylobacter* concentration was 50% of the mean *E. coli* concentration. The *E. coli* concentration variation was considered as an indicator of *Campylobacter* concentration variation. The *E. coli* variation was applied to the average *Campylobacter* concentration to improve the estimate of variation in the stochastic assessment.

**Table 2.4** Results of Monte Carlo simulations in approaches 2, 3 and 4

| | approach 2 | approach 3 | approach 4 |
|---|---|---|---|
| raw water MPN/L | 199 | 199 | 199 |
| filtration log reduction | 1.1 | 1.04 | 1.04 |
| ozonation log reduction | 1.6 | 1,74 | 1.58 |
| slow sand filtration log reduction | 2.0 | 2.0 | 2.0 |
| **Monte Carlo simulation results** | | | |
| total reduction | 4.6 | 4.7 | 4.6 |
| log pathogen concentration | −2.3 | −2.4 | −2.3 |
| 95% of log *Campylobacter* concentration | −1.7 | −1.9 | −1.8 |
| 99.7% 'worst day' log *Campylobacter* concentration | −0.6 | −0.8 | −0.5 |

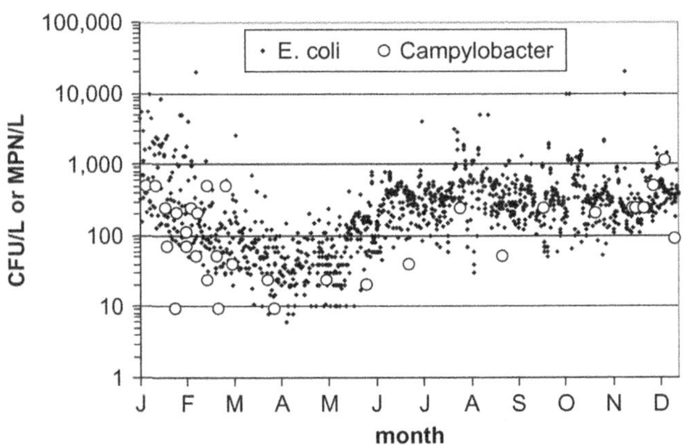


**Figure 2.7** Yearly variation of reported *E. coli* ('00–'05) and *Campylobacter* ('03–'05) concentrations in raw water

*Campylobacter* and *E. coli* are similarly reduced by rapid sand filtration and slow sand filtration (Hijnen *et al.*, 2005b) and ozonation (Smeets *et al.*, 2005). The *E. coli* reduction by filtration and ozonation was applied in the assessment as a surrogate of *Campylobacter* reduction. Table 2.5 shows the resulting values of the assessment. The uncertainty about the mean values of $C_{in}$, $\pi_{rsf}$ and $\pi_{O3}$ in Table 2.5 shows that the 95% confidence intervals of the fitted PDFs were very small due to the large amount of *E. coli* data. Results of the Monte Carlo simulation using the *E. coli* data were similar to the results of the simulation with *Campylobacter* data (Table 2.4). Although the use of surrogate data did not affect the estimate of pathogen concentrations in drinking water, it did reduce the uncertainty with respect to the appropriate PDF parameters.

## Approach 4: Treatment modelling

Approach 4 was similar to approach 3 with the exception that the efficacy of ozonation was assessed by modelling *Campylobacter* inactivation using monitored process conditions. This approach was an example of how different types of data could be combined in a treatment assessment. Figure 2.8 illustrates how the different types of data were used in each step.

**Table 2.5** Results of indicator assessment in approach 3. $\alpha$ and $\beta$ are the MLE parameters of the fitted PDF

| | lower 95% CI boundary | MLE | upper 95% CI boundary | $\alpha$ MLE | $\beta$ MLE | PDF type |
|---|---|---|---|---|---|---|
| raw water (95% CI) MPN/L | 174 | 199 | 226 | 0.78 | 256 | gamma |
| filtration log reduction | 1.06 | 1.04 | 1.02 | 0.55 | 5.41 | beta |
| ozonation log reduction | 1.78 | 1.74 | 1.70 | 0.42 | 22.8 | beta |
| slow sand filtration log reduction | * | 2.0 | * | 0.41 | 36 | beta |

**No parameters for the 95% CI boundaries of the PDF could be determined.*

The result of the process model in Table 2.6 shows that the calculated inactivation of *Campylobacter* by ozone was slightly lower than when microbial monitoring was used (Table 2.3 and Table 2.5). However, this had only a limited effect on the estimated *Campylobacter* concentrations in treated water. The modelled inactivation was much lower than the point estimate in approach 1 and more in line with the microbial assessments in approaches 2 and 3. This shows that the process settings used in approach 1 were not always achieved in practice. Moments of sub-optimal conditions were incorporated in the monitoring data used in approach 4 and as a result the modelled inactivation was lower.

**Table 2.6** Results of approach 4 where pathogen data, indicator data and process modelling and literature data were combined. $\alpha$ and $\beta$ are the MLE parameters of the fitted PDF

| | lower 95% CI boundary | MLE | upper 95% CI boundary | $\alpha$ MLE | $\beta$ MLE | PDF type |
|---|---|---|---|---|---|---|
| raw water (95% CI) MPN/L | 174 | 199 | 226 | 0.78 | 256 | gamma |
| filtration log reduction | 1.02 | 1.04 | 1.06 | 0.55 | 5.41 | beta |
| ozonation log reduction | 1.62 | 1.58 | 1.52 | 0.25 | 8.8 | beta |
| slow sand filtration log reduction | * | 2.0 | * | 0.41 | 36 | beta |

**No parameters for the 95% CI boundaries of the PDF could be determined.*

## DISCUSSION

### Point estimate versus stochastic assessment

Several approaches for QMRA were applied to a case study treatment system. The point estimate assessment required no site specific information and used a simple calculation method. The worst case point estimate of *Campylobacter* concentration in treated water was almost six orders of magnitude higher than the best estimate and 16 orders of magnitude higher than the minimum estimate. In the sensitivity analysis, 38% of the combinations led to *Campylobacter* concentrations exceeding the best estimate. The point estimate provided an indication of the possible *Campylobacter* concentration in drinking water at the treatment site. Since the point estimate combined raw water quality with the extent of treatment, it could be used to compare and prioritise different treatment systems. The point estimate showed that, as a result of extreme conditions, the *Campylobacter* concentration in drinking water could vary over 16 orders of magnitude. However, it provided no information about the likelihood of these extreme conditions. Therefore the point estimate may not provide sufficient basis for decisions by the risk manager or the legislator.

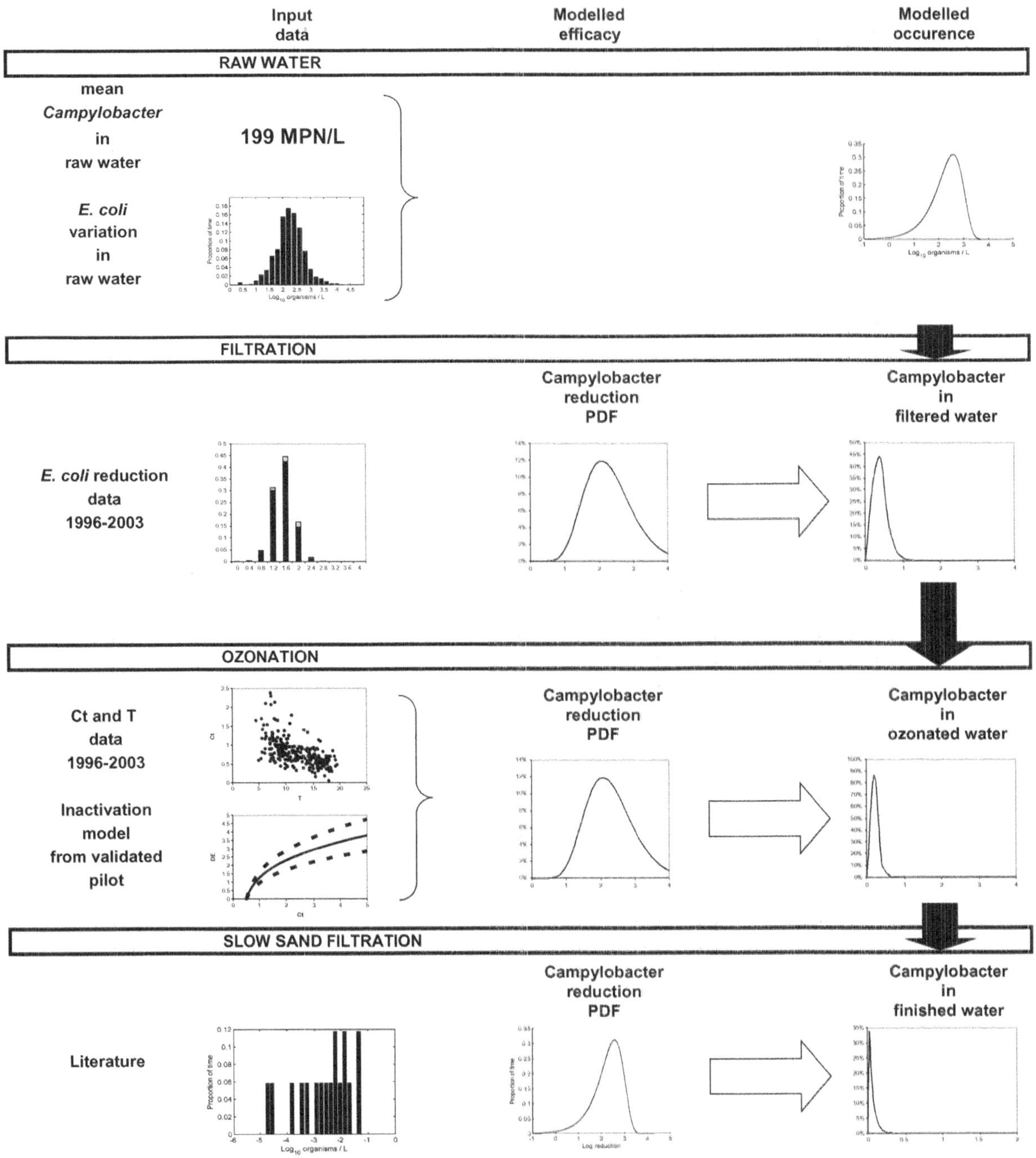


**Figure 2.8** Visualization of the model calculations for approach 4 where different types of site specific data were combined in a stochastic assessment

Approach 2 used *Campylobacter* monitoring data in a stochastic assessment. By fitting the MLE and the 95% CI PDF to the data, the uncertainty about the mean $C_{in}$ could be determined. Similarly the uncertainty with respect to the reductions $\pi_{rsf}$, $\pi_{O3}$ and $\pi_{ssf}$ was quantified. In addition, the stochastic assessment provided information about the variability of the system so the frequency and magnitude of peak concentrations could be estimated. The use of site specific information increased the credibility of the outcome of the stochastic assessment.

## Pathogen versus indicator organism

The uncertainty about the raw water PDF parameters was reduced from 0.8 logs (74–535 MPN/L) to 0.1 logs (174–226 MPN/L) in approach 3 by using *E. coli* monitoring data. The *E. coli* data included sixty times more data points than the *Campylobacter* data. Since the uncertainty about the PDF parameters was reduced, the uncertainty about the mean concentration was also reduced. However, the assumption that variation in source water and reduction by treatment of *Campylobacter* and *E. coli* were similar introduced some uncertainty that was not quantified. The estimated mean and peak *Campylobacter* concentrations were similar to the estimates in approach 2 since the *E. coli* and *Campylobacter* data showed similar variation. This illustrated that collecting more data by intensive monitoring could reduce uncertainty. However, intensive microbial monitoring may not be feasible in all situations.

## Use of process models

Modelling ozone efficacy as a point estimate based on operational setpoints in approach 1 resulted in an overestimation of inactivation compared to the assessment in approaches 2, 3 and 4. The overestimation of inactivation by ozone in approach 1 was caused by wrong assumptions on process conditions. Short periods of low ozone concentrations, observed in the data, were not accounted for by the point estimate. In microbial risk assessment, process models without detailed site specific information should also account for these suboptimal conditions.

Approach 4 showed how information of process operation, such as ozone concentration, contact time and temperature, could be used to assess treatment efficacy through a disinfection model. Process modelling resulted in estimates of mean and peak *Campylobacter* concentrations similar to approaches 2 and 3. This verified that the model provided a reliable estimate of inactivation by ozone. Process modelling could be applied when microorganism concentrations are to low to monitor. In addition, process modelling provides an opportunity to further improve the assessment of ozonation. By monitoring $C_{O3}$, $T$ and $t$ on-line and using the model, a very large dataset of ozonation efficacy, exceeding that of *E. coli*, could be acquired with a minimum of resources. Such a dataset could further reduce the uncertainty about ozonation efficacy.

The slow sand filtration was included using data from literature in all approaches. The differences in reported efficacies may be caused by variability of the process e.g. due to differences in design (sand grain, filterbed depth), operation (filtration rate, scraping) and conditions (temperature, organic load). The efficacy of slow sand filtration was described by a beta PDF. However, the PDF-fit method could not determine 95% CI parameters, and looking at Figure 2.2 a uniform distribution of log reductions might also be applicable. More site specific data on slow sand filtration could improve the assessment. Site specific data about slow sand filtration could be obtained by experiments with dosed microorganisms, since microorganism concentrations in the influent are generally very low. These experiments could support a choice for PDF type and reduce the uncertainty.

## Uncertainty of the assessment

Approaches 2, 3 and 4 assessed the uncertainty about the raw water *Campylobacter* concentration and the treatment efficacy. This was illustrated in Tables 2.3, 2.5 and 2.6 by the 95% CI of the mean concentration and efficacies. These uncertainties were not included in the Monte Carlo simulations. Assumptions such as the applicability of *E. coli* as a surrogate for *Campylobacter*, the choice of PDF type for slow sand filtration and the independence between treatment steps also introduce some uncertainty in the assessments. In addition the PDFs of $C_{raw}$, $\pi_{rsf}$, $\pi_{O3}$ and $\pi_{ssf}$ were based on data that generally included some level of uncertainty such as recovery, quantification of the microbial method (MPN method) and culturability and infectivity of the organisms. These uncertainties were not specifically addressed in the presented stochastic methods. However, they may have an impact on the estimated *Campylobacter* concentration in drinking water. When the stochastic assessments are used for decision support by risk managers or legislators, the uncertainty of the assessment results needs to be taken into account. Development of methods that specifically address uncertainty in the drinking water treatment assessment is therefore needed.

## CONCLUSIONS

A point estimate assessment was compared to three stochastic assessments of drinking water treatment efficacy. Although the point estimate assessment provided an indication of the *Campylobacter* concentrations in drinking water, it provided no information about the uncertainty of this estimate. Minimum and maximum estimates ranged over 16 orders of magnitude. The stochastic method using monitored *Campylobacter* data showed that the estimated concentrations in drinking water indeed varied over several orders of magnitude. The frequency and magnitude of peak concentrations and their impact on the pathogen breakthrough to treated water could be estimated. The stochastic methods also provided insight in the uncertainty of the estimated raw water concentrations and treatment efficacy. Increasing the amount of data by using monitored *E. coli* as a surrogate for *Campylobacter* reduced the uncertainty about the variability. When monitored process conditions were used to model inactivation by ozonation, the results were similar to the results of the microbial assessments. Process modelling could be applied when no microbial data was available and could allow for on-line monitoring of treatment efficacy. The presented stochastic methods focussed on including variability in the assessment. However, uncertainty about the elements of the assessment may have an impact on the result. Development of methods to include uncertainty in the stochastic assessment is therefore needed, especially when risk assessment results are used for decision support or legislation.

## REFERENCES


Anonymous (1996). NEN 6269 *Bacteriologisch Onderzoek Van Water – Onderzoek Naar De Aanwezigheid En/of Het Meest Waarschijnlijke Aantal Van Thermofiele Campylobacterbacteriën* (standardized method for analysis of *Campylobacter* in water) Nederlands Normalisatie Instituut, Delft, the Netherlands.

Bartram J., Fewtrell L. and Stenström T. (2001). Harmonised Assessment of Risk and Risk Management for Water-Related Infectious Disease: An Overview. In: *Water Quality: Guidelines, Standards and Health. Assessment of risk and risk management for water-related infectious disease*, Fewtrell L. and Bartram J. (eds.). IWA Publishing, London, UK.

De Man J. C. (1975). The probability of most probable numbers. *European Journal of Applied Microbiology*, **I**, 67–78.

Do-Quang Z., Ramirez C. C. and Roustan M. (2000). Influence of geometrical characteristics and operating conditions on the effectiveness of ozone contacting in fine-bubbles conventional diffusion reactors. *Ozone: Science and Engineering*, **22**, 369–378.

Haas C. N., Rose J. B. and Gerba C. P. (1999). *Quantitative Microbial Risk Assessment*. John Wiley & Sons, New York, USA.

Hijnen W. A. M., Beerendonk E. and Medema G. J. (2005a). *Elimination of Microorganisms by Drinking Water Treatment Processes (second edition)*. Kiwa NV, Nieuwegein, the Netherlands.

Hijnen W. A. M., Beerendonk E. and Medema G. J. (2005b). *Additional Memo Including Rapid Sand Filtration, GAC Filtration and Direct Filtration to Elimination of Microorganisms by Drinking Water Treatment Processes* Kiwa NV, Nieuwegein, the Netherlands.

Medema G. J. (2003). Integration of quantitative microbial risk assessment (QMRA) and risk management in water safety Plans. *Water safety conference Berlin 28–30 April 2003, Report of a workshop*, Kiwa, Nieuwegein, the Netherlands.

Regli S., Rose J. B., Haas C. N. and Gerba C. P. (1991). Modeling the risk from *Giardia* and viruses in drinking water. *Journal of the American Water Works Association JAWWA*, **83**(11), 76–84.

Smeets P. W. M. H., Dullemont Y. J. and Medema G. J. (2005). *E. coli* as surrogate for *Campylobacter* inactivation at bench-scale and full-scale and high ozone resistance of environmental *E. coli* and *Campylobacter*, *17th IOA conference 22–26 August 2005*, Strasbourg, France.

Smeets P. W. M. H., Van der Helm A. W. C., Dullemont Y. J., Rietveld L. C., van Dijk J. C. and Medema G. J. (2006). *E-coli* inactivation by ozone under bench-scale and full-scale hydraulic conditions. *Water Research*, **40**, 3239–3248.

Teunis P. F. M., Medema G. J., Kruidenier L. and Havelaar A. H. (1997). Assessment of the risk of infection by *Cryptosporidium* or *Giardia* in drinking water from a surface water source. *Water Research*, **31**(6), 1333–1346.

Teunis P. F. M., Evers E. G. and Slob W. (1999). Analysis of variable fractions resulting from microbial counts. *Quantitative Microbiology*, **1**, 63–88.

Teunis P. F. M., Davidson A., and Deere D. (2004). *Short Term Fluctuations in Drinking Water Quality and Their Significance for Public Health.* World health organization, Geneva, Switzerland.

USEPA (2003). LT2ESWTR *Long Term Second Enhanced Surface Water Treatment Rule and Draft Toolbox Guidance Manual*, USEPA, Washington DC, USA.

WHO (2004). *Guidelines for Drinking Water Quality (third edition)*. World Health Organization, Geneva, Switzerland.

# Chapter 3

# How can the UK statutory *Cryptosporidium* monitoring be used for quantitative risk assessment of *Cryptosporidium* in drinking water?

*P.W.M.H. Smeets[a,c], J.C. van Dijk[c], G. Stanfield[b], L.C. Rietveld[c] and G.J. Medema[a,c]*

[a]*KWR Watercycle research institute, PO Box 1072, 3430 BB Nieuwegein, the Netherlands*
[b]*WRC plc, Frankland Road, Blagrove, Swindon, Wiltshire SN5 8YF, UK*
[c]*Faculty of Civil Engineering, Delft University of Technology, PO Box 5048, 2600 GA Delft, the Netherlands*

## ABSTRACT

Quantitative Microbial Risk Assessment (QMRA) is increasingly being used to complement traditional verification of drinking water safety through the absence of indicator bacteria. However, the full benefit of QMRA is often not achieved because of a lack of appropriate data on the fate and behaviour of pathogens. In the UK, statutory monitoring for *Cryptosporidium* has provided a unique dataset of a specific pathogen directly measured in large volumes of treated drinking water. Using this data a QMRA was performed to determine the benefits and limitations of such state of the art monitoring for risk assessment. Estimates of the risk of infection at the 216 assessed treatment sites ranged from $10^{-6.5}$ to $10^{-2.5}$ person$^{-1}$.day$^{-1}$. In addition, *Cryptosporidium* monitoring data in source water was collected at eight treatment sites to determine how *Cryptosporidium* removal could be quantified for QMRA purposes. *Cryptosporidium* removal varied from 1.8 to 5.2 log units and appeared to be related to source water *Cryptosporidium* concentration. Application of general removal credits can either over- or underestimate *Cryptosporidium* removal by full-scale sedimentation and filtration. State of the art pathogen monitoring can identify poorly performing systems, although it is ineffective to verify drinking water safety to the level of $10^{-4}$ infections person$^{-1}$.year$^{-1}$.

 *Stochastic Modelling of Drinking Water Treatment in Quantitative Microbial Risk Assessment.* By Patrick Smeets. ISBN: 9781843393740. Published by IWA Publishing, London, UK.

## INTRODUCTION

Drinking water is generally monitored for the occurrence (absence) of indicator organisms in a relatively small volume of water (<500 ml), but absence in this volume does not guarantee safe drinking water. Additional approaches to safeguard drinking water quality, such as the Surface Water Treatment Rule (USEPA, 2006) and Water Safety Plans (WHO, 2004), have therefore been introduced. Quantitative Microbial Risk Assessment (QMRA) (Haas *et al.*, 1999) is a method that can be used to estimate the health risk associated with drinking water consumption. The risk of infection is calculated from the exposure to pathogens (the chance of ingesting one or more pathogens) and the dose-response relation (the chance of infection from the number of pathogens ingested). In most situations it is not feasible to monitor directly for the presence of pathogens since they are present in extremely low numbers. However, the UK statutory monitoring for *Cryptosporidium* has resulted in a large quantity of data from the monitoring of large volumes of drinking water. The UK Water Supply Regulations (Amended) 1999 (DWI, 1999), that were introduced as a treatment rather than a health-based standard, require continuous monitoring of finished drinking water for *Cryptosporidium* oocysts at least 23 hours per day at a rate of at least 40 litres per hour. This could be considered the best possible pathogen monitoring programme that can be achieved given the current state of the art of pathogen analysis. The database of monitoring results provides a unique opportunity to study drinking water safety based on measured pathogens in drinking water. The monitoring is performed to ensure compliance with the UK drinking water treatment standard of less than 1 oocyst per 10 L of final water. The endpoint of QMRA is an estimated risk of infection, and Dutch drinking water regulations require a maximum risk level as low as $10^{-4}$ per person per year. This study investigated whether extensive monitoring of treated water as applied in the UK could be used to verify compliance with this risk level.

Since most pathogens are not monitored as extensively in drinking water, a different approach is generally used in QMRA. Pathogens are monitored in source water, and their reduction through treatment is estimated so the concentration in drinking water can be calculated (Regli *et al.*, 1991). An essential step in such a risk assessment is determining the reduction of pathogens by drinking water treatment (Teunis *et al.*, 1997 and Gibson *et al.*, 1999). Generally, reduction has to be estimated from indirect information such as; process removal credits from literature or pilot experiments, removal of surrogates like turbidity, reduction of indicator organisms such as *E. coli* or modelling of disinfection processes. These indirect measurements of pathogen reduction all have their specific shortcomings. Direct measurement of pathogens before and after treatment could provide a direct measurement of treatment efficacy. For some sites *Cryptosporidium* monitoring data was collected for the source water to see if this would aid the modelling of treatment in QMRA.

Changes in conditions that are outside the limits normally experienced at a treatment works are often referred to as extreme events. The most common extreme event that is considered is heavy rainfall (storms) in the catchment of the treatment works. Abnormally high rainfall can wash pathogens from agricultural land used for grazing into the river (the source water) (Signor *et al.*, 2007). If there has been little or no rainfall previously, the increased numbers of pathogens scoured from the land is not offset by the increased dilution and, at least initially, the numbers of pathogens in the source water will increase. The flow rate of the river also increases, causing high turbidity and carrying pathogens more quickly to the abstraction point of the water treatment works. The concern is whether the treatment works can handle this increased microbial challenge or if the quality of the treated water will deteriorate in relation to the increased loading in the source water and how this can be modelled in QMRA.

The goal of this study was to test whether elaborate end-product testing can provide valuable information for quantitative risk assessment. On the one hand risk of infection was estimated directly from the treated water monitoring. On the other hand *Cryptosporidium* reduction by treatment was assessed by comparing *Cryptosporidium* in the source and treated water. Finally the impact of peak events in the source water on treated water quality was studied.

## METHODS

For this study, the results of statutory *Cryptosporidium* monitoring at 216 UK water treatment works from 1/4/2000 until 31/3/2002 were obtained from WRC plc, Swindon, UK. The daily sampled volume and the oocysts concentration in that volume were used in the data analysis. In addition data from source water monitoring at 8 UK water treatment works (Sites A-H) from 5-1-1993 to 20-4-2004 was obtained. These samples were taken at irregular intervals, typically between one week and one month. The treatment schemes at these sites are presented in Table 3.1. All sites apply Coagulation, Sedimentation, Filtration, GAC filtration and Chlorination. At sites A and E ozonation is also used and at site B there is also Dissolved Air Flotation.

To study the effect of source water peaks one UK Water Company (not described in Table 3.1) submitted *Cryptosporidium* monitoring data from one of its water treatment works that was fortuitously collected during a period of extreme rainfall. At this treatment works, the river water is abstracted for treatment. The first stage of treatment consists of bank-side storage of around 2 days. The initial coagulation stage is a process that uses sand and polyelectrolyte with aluminium sulphate. After settlement the water is passed through a bank of rapid gravity filters. Ozone is added and after contact the water passes through GAC filters before final chlorination. Statutory monitoring for *Cryptosporidium* is carried out on the final water from this works. During the period, immediately before and for two weeks after, additional source water monitoring was carried out in the form of one 10 litre grab sample each day.

**Table 3.1** Treatment processes at sites A to H

| Site | Treatment |
|---|---|
| A | Coagulation Sedimentation GAC filtration Ozone Chlorination |
| B | Impoundment Coagulation (Polyelectrolyte) Sedimentation Dissolved Air Flotation Filtration GAC filtration Chlorination |
| C | Coagulation (Polyelectrolyte) Sedimentation Filtration GAC filtration Chlorination |
| D | Impoundment Coagulation (Polyelectrolyte) Sedimentation Filtration GAC filtration Chlorination |
| E | Coagulation (Polyelectrolyte) Sedimentation Filtration GAC filtration Ozone Chlorination |
| F | Coagulation (Polyelectrolyte) Sedimentation Filtration GAC filtration Chlorination |
| G | Coagulation (Polyelectrolyte) Sedimentation Filtration GAC filtration Chlorination |
| H | Coagulation Sedimentation Filtration GAC filtration Chlorination |

The methods used for *Cryptosporidium* monitoring and analysis in drinking water in connection with the UK Water Supply Regulations (Amended) 1999 are strictly controlled by the UK Drinking Water Inspectorate (DWI). The methods are detailed in supplements to the Regulations and include a requirement to provide a high level of security such that a 'chain of evidence' is produced to allow the results to be admissible in a court of law. The DWI stipulated methods include Sampling and Transportation of Samples (Part 1), Laboratory and Analytical procedures (Part 2), Validation of new Methods (Part 3) and Requirements for the Inter-laboratory Proficiency Scheme (Part 4). All laboratories undertaking *Cryptosporidium* analysis in connection with the regulations must take part in the inter laboratory scheme. In addition the laboratories are regularly inspected by inspectors from the DWI and are awarded a license of proficiency that can be revoked at subsequent inspections if standards have fallen below the required standard. The regulations are given in full on the DWI Internet site (www.dwi.gov.uk). Basically the method of sampling and analysis involved filtration at the sampling point using an IDEXX Filter Max© filter or a Pall Life Sciences Envirochek™ filter. On return to the laboratory the filters go through an elution process. The eluent is then treated with centrifugation and immuno-magnetic separation to concentrate the oocysts. The concentrate is then treated with an immuno-fluorescent reagent and examined microscopically. The requirements stipulated in the DWI Standard Operating Procedures (SOP) are extremely strict about each of the stages of the analyses in an attempt to retain the 'chain of custody', maximise the efficiency of recovery and provide the best means of comparing results from different laboratories. For example the SOP for microscopic examination requires that where an initial result of 0.5 oocysts per 10 litres is obtained, the microscopic examination should be checked by another approved microscopist in the same laboratory. If the results of initial analysis suggests more then 0.7 oocysts per 10 litres or there is doubt about identification of some 'oocysts', the microscopic examination should be checked by one of the DWI approved microscopists from another organisation as well as being examined by another microscopist in the same laboratory. The recovery of the method is approximately 30%–60%, recovery is not determined for individual samples (DWI, 2006). On the one hand such recovery leads to an underestimation of *Cryptosporidium* concentrations, on the other hand only part of the counted oocysts are human pathogenic, viable and infectious (Aboytes *et al.*, 2004). Since there is insufficient data to quantify these effects, it was assumed that the net effect on the assessment was negligible.

Oocyst concentrations were plotted in a Complementary Cumulative Distribution Function (CCDF) graph of exceeded concentration on a log-log scale. Non-detects are not shown but determine the starting point of the graph. To determine the impact of possible oocysts concentrations represented by the non-detects, three approaches were used to extrapolate the measured oocyst concentrations to below detection limits. For the minimal estimated risk of infection, these concentrations were set to 0, assuming that no oocysts were present in the drinking water produced on that day. For the maximum estimate, these samples were set to the detection limit (1 oocyst per 1000 L). For the best estimate the concentration in the non-detect samples was extrapolated linearly on a log-log scale (log-linear) as:

$$\log_{10}(frequency) = a + b \bullet \log_{10}(C_{crypto}) \tag{1}$$

Where $C_{crypto}$ is the *Cryptosporidium* concentration in oocysts.$L^{-1}$. Parameters a and b were determined from the lowest 10 measured concentrations. If the number of positive samples was between 2 and 10, all samples were used for the extrapolation. Single or no positives were not extrapolated.

Source water was generally monitored in 1 to 10 L samples. Recovery in the source water can be influenced by water quality changes such as high turbidity (Nieminsky *et al.*, 1995 and Wohlsen *et al.*, 2004). In this study the recovery for all samples was assumed equal since the recovery was not assessed for individual samples. The reduction of *Cryptosporidium* at the assessed treatment sites A to H relies on the physical removal by sedimentation and filtration. Chlorination has a very limited effect on *Cryptosporidium* under normal operating conditions. Site A and E apply ozonation which could inactivate oocysts. Since the inactivated oocysts are not removed from the water, they are counted by the analysis. Therefore this assessment only evaluated the physical removal of *Cryptosporidium* by treatment. Average concentrations before and after treatment at each site were calculated as the total number of oocysts counted in the total sample volume. Log removal was calculated from the average concentrations. The number of *Cryptosporidium* in the source water and the total volume of the treated water samples determine the maximum level of reduction that can be demonstrated at a site. The demonstrable *Cryptosporidium* reduction by treatment works was calculated by assuming that only one *Cryptosporidium* was counted in the total treated water sample volume.

Risk of infection was calculated both as a point estimate based on average values and as a varying risk using Monte Carlo simulation. Treated water *Cryptosporidium* concentrations were determined from the data. Drinking water consumption in the UK was modelled with a Poisson distribution for the number of 190 ml glasses per day (mean of 2.81 glasses) (Mons *et al.*, 2006). For the point estimate the mean consumption of

0.53 L was applied. A Beta-Poisson dose-response model ($\alpha = 0.115$, $\beta = 0.176$) was used to calculate the risk of infection when exposed to a concentration of *Cryptosporidium* (Teunis *et al.*, 2002):

$$P_{inf\_d} = 1 - \left( \left( 1 + \frac{C_{crypto} \bullet consumption}{\beta} \right) \right)^{-\alpha} \quad (2)$$

$P_{inf\_d}$ is the daily risk of infection ($person^{-1}.day^{-1}$), and *consumption* is the volume of daily consumed unboiled drinking water (L). The yearly risk of infection $P_{inf\_y}$ ($person^{-1}.year^{-1}$) is generally approximated by 365 times the daily risk of infection. However, at daily risks above $10^{-3.5}$ ($person^{-1}.day^{-1}$) this approach significantly overestimates the actual risk. Therefore the yearly risk of infection was calculated as (Haas *et al.*, 1999):

$$P_{inf\_y} = 1 - (1 - P_{inf\_d})^{365} \quad (3)$$

The Monte Carlo simulation was performed by 100,000 independent draws from monitored *Cryptosporidium* concentrations and the applicable type of extrapolation.

# RESULTS AND DISCUSSION

## Overview of monitoring results

The treated water results of 216 sites, including sites A to H, were collected. Table 3.2 provides an overview of these results.

Figure 3.1 shows the frequency of observed concentrations is almost linear on a log-log scale. The minimal, maximal and log-linear extrapolation of negative sample results are also represented in the graph. For each concentration the exposure to *Cryptosporidium* ($C_{crypto}$*frequency) was calculated to show the relative impact of the observed concentrations. The exposure at relatively low concentrations between 0.001 and 0.01 $oocysts.L^{-1}$ seems to be five times higher than the exposure at high concentrations between 0.5 and 1 $oocyst.L^{-1}$, since the latter rarely occur.

For site A to H 4,214 samples of source water and 5,579 samples of finished water, representing 133,500 and 6,826,549 L of tested water respectively were collected. Each of the 8 sites had analysed approximately 850,000 L of finished water in 700 samples of 1,200 L over a period of 23 months. Table 3.3 provides an overview of the monitoring results.

**Table 3.2** Overview of statutory UK *Cryptosporidium* monitoring results

| | | |
|---|---|---|
| Number of sites | 216 | |
| Number of samples | 97,997 | |
| # Positive | 5353 | (5.5%) |
| Total sample volume | 115,303,050 | L |
| Total # oocysts counted | 24919 | |
| Average oocysts concentration | 0.000216 | $oocysts.L^{-1}$ |
| # Non-compliance (>0.1 $oocysts.L^{-1}$) | 18 | (0.018%) |

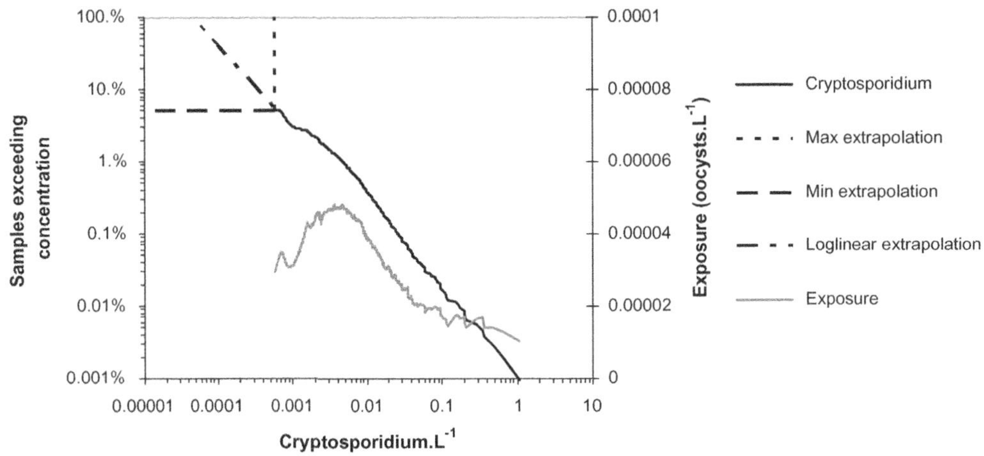


**Figure 3.1** Complementary Cumulative Frequency Distribution (CCDF) of observed *Cryptosporidium* concentrations (black), exposure due to these concentrations (grey) and three approaches to extrapolate the 94.5% negative samples (dashed)

**Table 3.3** Results of source water and finished water monitoring at site A to H

| | Source water | | | | Treated water | | | |
|---|---|---|---|---|---|---|---|---|
| | # samples | % positive | oocysts counted | volume analysed (L) | # samples | % positive | oocysts counted | volume analysed (L) |
| A | 470 | 2% | 8 | 2,746 | 702 | 0.57% | 4 | 817,989 |
| B | 643 | 18% | 179 | 3,561 | 691 | 0.43% | 6 | 828,326 |
| C | 489 | 43% | 701 | 1,289 | 710 | 0.42% | 3 | 820,731 |
| D | 1025 | 24% | 275 | 3,756 | 698 | 0.29% | 2 | 863,874 |
| E | 25 | 44% | 36 | 292 | 699 | 0.72% | 5 | 948,163 |
| F | 467 | 20% | 99 | 2,083 | 698 | 0.86% | 6 | 834,655 |
| G | 397 | 39% | 191 | 1,335 | 695 | 2.01% | 14 | 889,946 |
| H | 698 | 31% | 332 | 118,473 | 686 | 2.92% | 37 | 822,861 |

The period of source water monitoring varied per site and did not (completely) overlap with the study period of finished water monitoring. Table 3.4 compares the overlapping source water monitoring period to all source water monitoring. All the source water data was used in the analysis of treatment efficacy, since it provided a better representation of source water concentration and variation than the limited number of samples in the overlapping period.

Seasons may have an impact on both the concentrations of *Cryptosporidium* in the source water and treatment efficacy. Between 25% to 50% of the source water samples were positive for *Cryptosporidium*. The positive monitoring results for all sites were plotted against the day of the year in Figure 3.2. The only seasonality observed is a decrease of *Cryptosporidium* in August and a slight increase of positives from September to December for most sites.

**Table 3.4** Average and maximum *Cryptosporidium* concentration (oocysts/L) in source water during treated water study period and all source water monitoring

| | Source water monitoring during treated water study period | | | All source water monitoring results | | |
|---|---|---|---|---|---|---|
| | # samples | average concentration | maximum concentration | # samples | average concentration | maximum concentration |
| A | 16 | 0.0008 | 0.10 | 470 | 0.0029 | 0.57 |
| B | 7 | 0.0592 | 1.54 | 643 | 0.0503 | 8.48 |
| C | 2 | – | – | 489 | 0.5438 | 420.00 |
| D | 28 | 0.0158 | 0.42 | 1025 | 0.0732 | 3.33 |
| E | 4 | 0.3095 | 1.20 | 25 | 0.1233 | 1.20 |
| F | 13 | 0.0021 | 0.40 | 467 | 0.0475 | 3.21 |
| G | 15 | 0.0567 | 1.00 | 397 | 0.1430 | 4.29 |
| H | 119 | 0.0021 | 1.00 | 698 | 0.0028 | 4.76 |

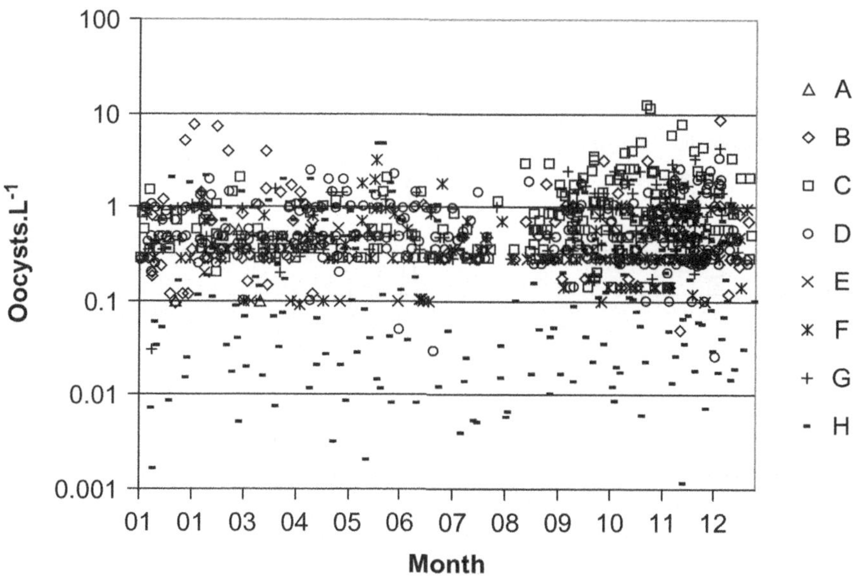


**Figure 3.2** Yearly observations of *Cryptosporidium* in source water (non-detects are not shown)

Seasonal variations like temperature or algal blooms could lead to compromised treatment resulting in *Cryptosporidium* in the drinking water. *Cryptosporidium* was detected in 5.5% of all treated water samples. Figure 3.3 shows how these samples were spread over the year. From January to March relatively little *Cryptosporidium* were detected, whereas the number of positive samples doubled in July.

The source water event data collected for a single site during the period before and after the rainfall are shown in Figure 3.4. The only detection of *Cryptosporidium* in the final water occurred on day 32 when one oocyst was detected in approximately 1,200 L.

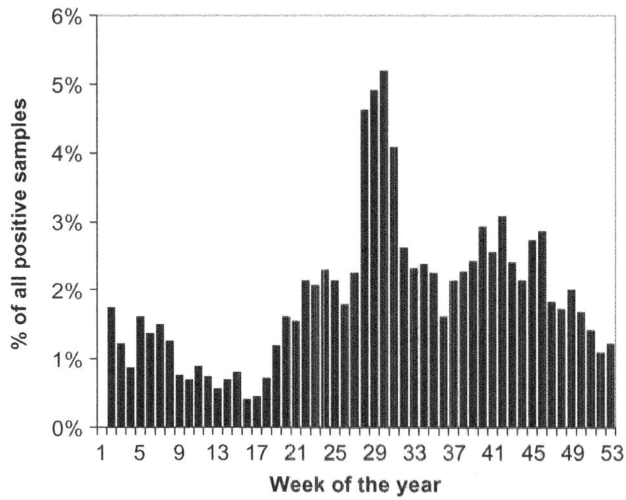


**Figure 3.3** Yearly distribution of all treated water samples containing *Cryptosporidium*

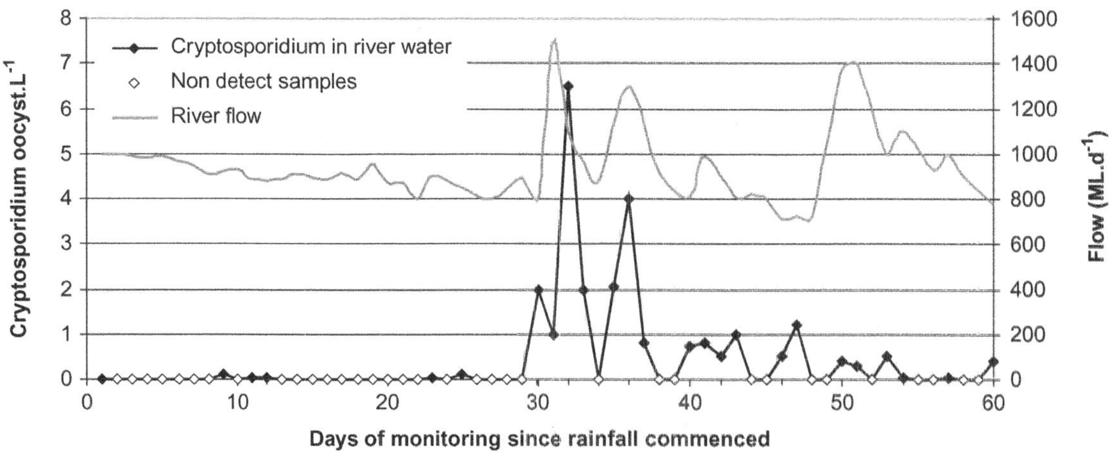


**Figure 3.4** Observation of peak contamination of *Cryptosporidium* in river water in relation to flow. A single oocyst was detected in a 1,200 L treated water sample on day 32 (not shown)

## QMRA based on treated water monitoring

The risk from the combined monitoring results of all sites was assessed under the assumption that each site produces the same amount of drinking water. The mean $P_{inf\_d}$ in Table 3.5 is the average risk of infection determined by Monte Carlo simulation. The point $P_{inf\_d}$ is the point estimate of average risk based on mean *Cryptosporidium* concentration, drinking water consumption and dose-response. The point estimate results in a slightly higher estimation of risk for all approaches. The estimated risk varies by a factor of 4 (0.6 log) depending on the approach for extrapolation of measured *Cryptosporidium* concentrations below the detection limit. The difference between log-linear and minimal approach is less than 0.03 log. The uncertainty caused by the negative samples has very little impact on this risk assessment. The yearly average risk of infection was $2.8*10^{-2}$ $person^{-1}.year^{-1}$.

The risk of infection was also assessed for individual sites based on their monitoring results. As an example, Figure 3.5 shows the distributions of *Cryptosporidium* concentrations in treated water for sites 22 to 28. Several typical distribution forms were observed. Sites 24 and 25 show a typical distribution where over 10% of the samples is positive. The frequency of occurrence decreases with increasing concentrations. Site 23 has a similar distribution, but is susceptible to peak events, shown by a higher increase of observed concentrations at decreasing frequencies. Site 22 is an example of a site with highly variable concentrations. Sites 27 and 28 show less variation at 1% and 10% positive samples respectively. Finally at site 26 only one oocysts was found in one sample, so it could not be extrapolated.

**Table 3.5** Daily risk of infection for the combined 216 UK sites included in the study

| Extrapolation | Monte Carlo simulation of mean $P_{inf\_d}$ | Point estimate of $P_{inf\_d}$ |
|---|---|---|
| Minimal | $7.45*10^{-5}$ | $8.08*10^{-5}$ |
| Loglinear | $7.98*10^{-5}$ | $8.75*10^{-5}$ |
| Max | $26.96*10^{-5}$ | $28.20*10^{-5}$ |

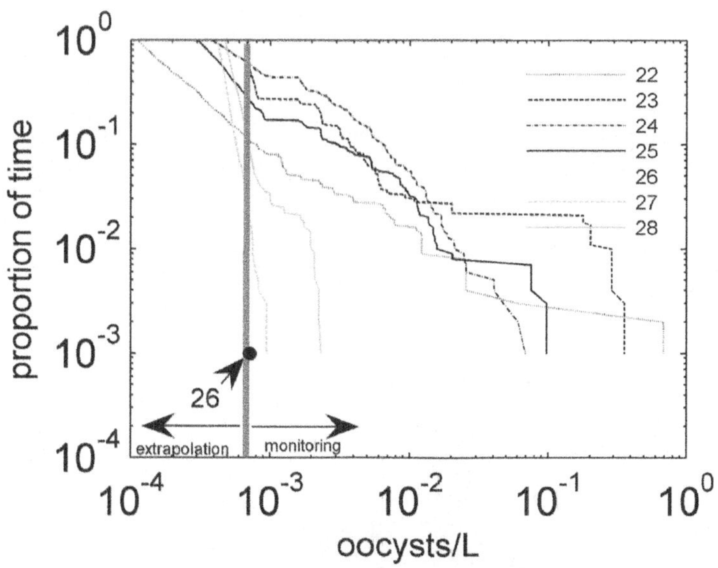


**Figure 3.5** Monitored *Cryptosporidium* concentrations in treated water at sites 22 to 28 including extrapolation by Monte Carlo simulation

The risk of infection was estimated with Monte Carlo simulation using the three different approaches to deal with non-detect samples. Figure 3.5 shows the resulting frequency of concentrations for the log-linear extrapolation.

The assessed daily risks for these sites are presented in Table 3.6. Since the minimal assessment of risk at sites 22, 23, 24 and 25 is relatively high it is hardly impacted by the way the negative samples are interpreted. At sites 27 and 28 extrapolation of the measured *Cryptosporidium* concentrations increases the assessed risk by one or two log units. The few 'extra' oocysts from the extrapolation have a strong impact on the low average concentration at these sites. Secondly, the few positive samples are all very low and show little variation.

When this little variation is extrapolated below detection limit, this results in a large percentage of the water being just below detection limit (steep extrapolation in Figure 3.5). Site 26 shows that a single detected oocyst is sufficient to achieve an average yearly risk of infection exceeding $10^{-4}$. When no *Cryptosporidium* is detected, assumptions on how to interpret these negative samples can still yield a risk exceeding $10^{-4}$ $person^{-1}.year^{-1}$ (or $2.7*10^{-7}$ $person^{-1}.day^{-1}$).

The frequency of minimal estimates of daily risk of infection at all 187 sites where *Cryptosporidium* was found (Figure 3.6) ranges from $10^{-6.5}$ to $10^{-2.5}$ $person^{-1}.day^{-1}$. The minimal risk estimate for the 29 sites where no *Cryptosporidium* was detected could be reported as $<5*10^{-7}$ $person^{-1}.day^{-1}$and the maximum risk estimate is approximately $3.4*10^{-4}$ $person^{-1}.day^{-1}$.

Risk levels were equally spread among sites with different sources (river, spring, groundwater or undefined), although the highest risks ($10^{-3.5}$ to $10^{-2.7}$ $person^{-1}.day^{-1}$) were not observed for groundwater and springs. There was no correlation between treatment throughput, number of samples or total monitoring volume and the risk of infection.

**Table 3.6** Risk assessment results for 7 individual sites

| site number | # of samples | % positive | Daily risk | | | Yearly risk |
|---|---|---|---|---|---|---|
| | | | minimal | loglinear | max | minimal |
| 22 | 417 | 7.7% | $3.8*10^{-4}$ | $4.25*10^{-4}$ | $8.06*10^{-4}$ | $1.5*10^{-1}$ |
| 23 | 201 | 53.7% | $1.6*10^{-3}$ | $1.60*10^{-3}$ | $1.73*10^{-3}$ | $4.4*10^{-1}$ |
| 24 | 453 | 53.9% | $9.6*10^{-4}$ | $1.01*10^{-3}$ | $1.08*10^{-3}$ | $2.9*10^{-1}$ |
| 25 | 459 | 29.0% | $4.8*10^{-4}$ | $6.20*10^{-4}$ | $7.59*10^{-4}$ | $1.9*10^{-1}$ |
| 26 | 454 | 0.2% | $4.7*10^{-7}$ | | $3.43*10^{-4}$ | $1.7*10^{-4}$ |
| 27 | 454 | 0.9% | $2.6*10^{-6}$ | $1.67*10^{-4}$ | $3.44*10^{-4}$ | $9.3*10^{-4}$ |
| 28 | 454 | 7.3% | $3.1*10^{-5}$ | $2.07*10^{-4}$ | $3.50*10^{-4}$ | $1.1*10^{-2}$ |

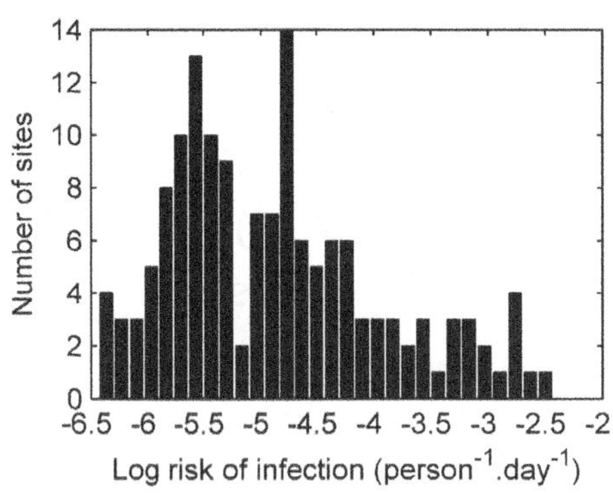


**Figure 3.6** Frequency of minimal estimates of log of risk of infection at the 187 sites where *Cryptosporidium* was found. At 29 sites no *Cryptosporidium* was detected

## Cryptosporidium removal

Table 3.7 and Figure 3.7 provide an overview of the over-all *Cryptosporidium* removal at sites A to H. The average treated water concentration is very similar for most sites, so the observed log removal is dominated by the source water concentration. Due to the small number of oocysts found in treated water, the assessed log reduction is slightly less than the demonstrable log reduction for most sites.

**Table 3.7** Over-all log removal at the treatment sites and demonstrable log removal

| site | average source concentration | average treated concentration | log removal | demonstrable log removal |
|---|---|---|---|---|
| A | 0.0029 | 0.0000049 | 2.78 | 3.38 |
| B | 0.0503 | 0.0000072 | 3.84 | 4.62 |
| C | 0.5438 | 0.0000037 | 5.17 | 5.65 |
| D | 0.0732 | 0.0000023 | 4.50 | 4.80 |
| E | 0.1233 | 0.0000053 | 4.37 | 5.07 |
| F | 0.0475 | 0.0000072 | 3.82 | 4.60 |
| G | 0.1430 | 0.0000157 | 3.96 | 5.10 |
| H | 0.0028 | 0.0000450 | 1.79 | 3.36 |

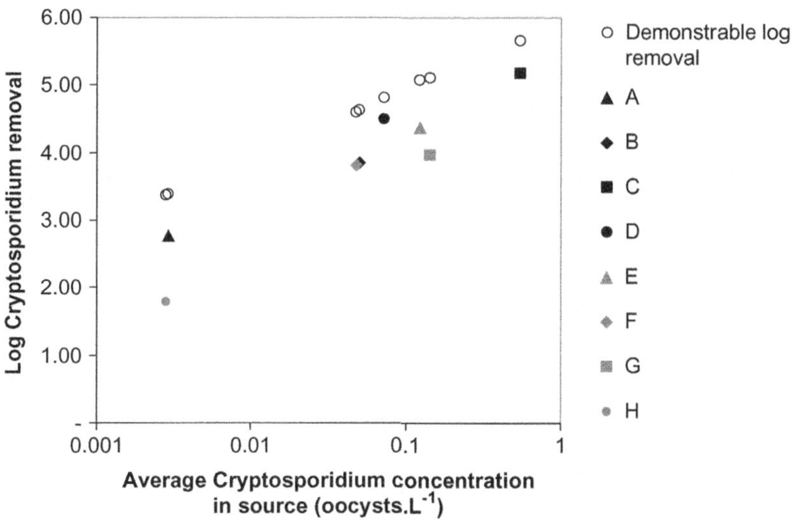


**Figure 3.7** Log reduction based on *Cryptosporidium* monitoring and demonstrable log reduction (if no *Cryptosporidium* had been found) at eight drinking water treatment sites in the UK

Since the physical treatment processes at sites A, C, D, E, F, G and H are of the same type one would expect similar log removal. A literature study resulted in a best estimate of 3.2 log removal of *Cryptosporidium* by conventional treatment (sedimentation, filtration) with a range of 1.4 to 5.5 log (Hijnen *et al.*, 2005b). Indeed 1.8 to 5.2 log removal was determined in this study. Site B could have provided more removal with the additional Dissolved Air Flotation, however this was not observed. Since treated water concentrations are similar at all sites, the removal appears to be related to the source water concentration. False positives or contamination in the laboratory could lead to such an observation. However this seems extremely unlikely for the analysis of *Cryptosporidium* given the strictness of the analysis procedures and the quality assurance regime demanded by the DWI. The following explanations for the differences in observed removal at these sites are discussed:

- The treatment design and operation is tailored to the water treated to comply with the drinking water standard.
- *Cryptosporidium* is accumulated in the treatment up to a steady state where sporadically some *Cryptosporidium* are released, independent of the source water concentration.
- Short high peaks of *Cryptosporidium* in the source are the cause of the positive samples in treated water. These peaks are not demonstrated by the infrequent monitoring of source water but are found in the continuous treated water sampling.
- Short 'failure' of treatment occurs at a similar frequency at all sites.
- Sedimentation and filtration are more effective to remove high concentrations of microorganisms than low concentrations.

## Design and operation

Sedimentation is generally optimised for the local situation. Regular jar tests are performed to determine optimal dosing of coagulant and coagulant aid and rapid and slow mixing energy. The removal of suspended matter or turbidity is thus optimised to reach a low turbidity after sedimentation. Polluted source water with high turbidity requires a better and more frequent optimisation than clean source water with low turbidity. The better optimised system also provides more *Cryptosporidium* removal, resulting in a relationship between source pollution and *Cryptosporidium* removal. Dugan *et al.* (2001) found that optimisation of sedimentation can improve *Cryptosporidium* removal from less than two to more than 5 log. In addition a filter receiving higher turbidity will show breakthrough of turbidity after the filter, so the filter can be backwashed in time. When turbidity is low before filtration, breakthrough may go unnoticed. Finally both sedimentation and filtration perform optimally within a range of suspended solids concentrations. Insufficient material can be present to form suitable flocks for sedimentation, while ripening of filters also requires sufficient suspended matter. Thus higher pollution drives treatments to improve their performance to meet both *Cryptosporidium* and other standards.

## Accumulation

Filtration processes capture particles (including *Cryptosporidium*) in the filter bed. Periodically the filter is backwashed to remove accumulated particles when the head loss over the filter has increased too much or when breakthrough of turbidity is observed. There is no direct relation between particles and *Cryptosporidium* concentration in the source water. So the amount of *Cryptosporidium* accumulated in the filter bed during a filter run is not directly related to the source water quality. Due to change of flow, some of the accumulated *Cryptosporidium* may detach from the filter material and leave the filter with the filtrate. Thus the number of *Cryptosporidium* after a filter could be more related to the loading of the filter than to the actual source water concentration. In the UK a system referred to as 'slow start up' is being used by water companies to bring a filter back on line after backwashing. This procedure has been shown to reduce the risk of residual oocysts in the filter being washed into the filtrate as the filter compacts after backwashing (WHO, 2004). Still some form of attachment and release might occur at a low level.

### Peaks in source water

The data for sites A to H were analysed to determine whether a recorded peak concentration in the source water had led to *Cryptosporidium* in the treated water. The source and treated water samples were combined by date. This showed that 57 treated water samples were positive, but unfortunately source water samples on the same date were only available on 5 occasions and 4 of these were negative for *Cryptosporidium*. The remaining single positive sample did not show a particularly high concentration. Neither were peaks in the source water observed in the periods preceding the 57 positive treated water samples. Comparing the average source water concentration to the maximum concentration in Table 3.4 shows that peaks of 10 to 200 times the average concentration have been recorded. So although these peaks could potentially lead to peaks in the treated water, this could not be confirmed from the UK statutory monitoring data.

The reported source water peak event in Figure 3.4 was studied in detail. A single oocyst was detected in the treated water on day 32. Because of the bank-side storage for 2 days the 'paired' source water value was a no detect in 10 L on day 29. However, due to mixing or preferential flow in the bank-side storage, the peak of 65 oocysts in 10 litres in the river water on day 32 may have led to an increased concentration at the intake of the treatment works on the same day. The fact that none of the treated water samples on the subsequent days 33–40 was positive for *Cryptosporidium* implies that the single detected oocyst was not related to the peak in the source water. These findings suggest that the detection of oocysts in treated water is not always related to peaks in source waters.

### Short treatment failure

The failure of equipment (e.g. a dosing pump, valve), installations (e.g. defective filter nozzle) or erroneous operation leading to decreased treatment performance is referred to as treatment failure. The occurrence of treatment failure is related to equipment age, maintenance and operational procedures. For the studied sites the frequency at which treatment failure occurs could be similar. However, if failure of treatment occurred at the same frequency at sites with different source concentrations this would lead to higher peaks at the more polluted sites. These sites only found low concentrations of one or two oocysts per sample, just like the less polluted sites. Therefore it is not likely that the similar occurrence of *Cryptosporidium* in the treated water of the studied sites is a consequence of treatment failure.

### Reduction related to microbial density

Some studies have observed that at high concentrations of microorganisms before slow sand filtration, more removal was found (Hijnen *et al.*, 2005a and Hijnen *et al.*, 2006). Removal of spores of sulphite-reducing Clostridia ranged over three log units at full-scale, and in some cases the concentration after filtration exceeded the influent concentration. They concluded that the high DEC-values assessed during short term dosing experiments most likely are not predictive for full-scale conditions. They attributed this observed relation between microorganism concentration and its removal to accumulation and release in the filter, as explained above.

## Modelling treatment in QMRA

In QMRA removal by treatment is modelled as a 'removal-credit' or by a distribution of removal values. According to the LTSESWTR (USEPA, 2003) a conventional treatment (coagulation, sedimentation and filtration) provides an average of 3 log *Cryptosporidium* removal when the treatment complies with the rule. The point estimates of removal at sites A to H show that in practice 1.8 to 5.2 log removal is achieved. So the type of treatment provides insufficient information to determine appropriate removal value(s) for *Cryptosporidium*.

Local information can verify substantial higher removal, leading to a lower risk estimate. Information about removal of turbidity, particles or surrogate organisms could support a choice of removal value, although this could not be verified in this study. The distribution of *Cryptosporidium* in treated water might be related to its distribution in source water. Therefore the frequencies of *Cryptosporidium* in source and treated water were combined in one figure for each site. In the example in Figure 3.8, treated water at site G shows less variation that at H. The source water varies more at site H. For these sites, a single removal-credit could be used to model removal by treatment in QMRA. However, these were the poorest performing sites. Since the number of positive samples after treatment was very low for sites A to F, variability could not be quantified. Since source water events do not lead directly to *Cryptosporidium* in the treated water, a single removal credit is not appropriate for these sites. This leaves the question on how to model sedimentation-filtration in QMRA. One option is to relate the removal to the source water concentration. This leads to little variation in the treated water. However, both large and little variation in treated water was observed in this study. In order to improve knowledge about treatment efficacy this study could be expanded with additional information, such as design characteristics, additional source water data from other sites and on-line measurement of surrogates (turbidity) to match *Cryptosporidium* monitoring. Such a full-scale study at many sites could then lead to effective models of treatment efficacy for QMRA.

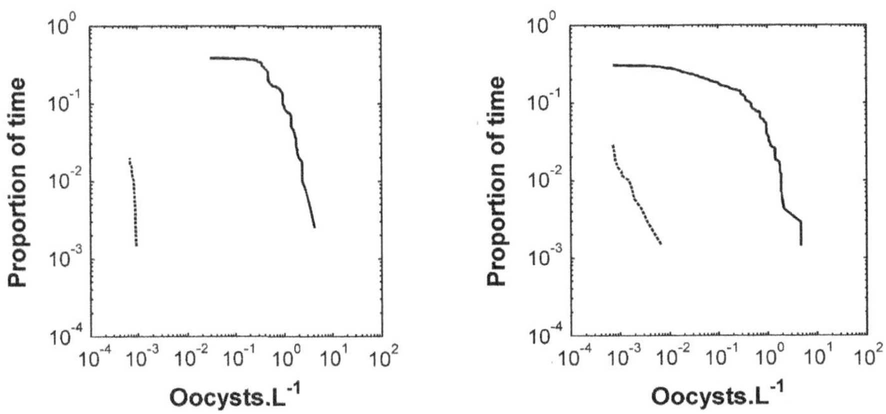


**Figure 3.8** Reported source water (line) and treated water (dashed) *Cryptosporidium* concentrations for sites G (left) and H (right)

## CONCLUSIONS

State of the art *Cryptosporidium* monitoring as required in the UK can be used to identify effectively those water supplies that do not reach a health-based target below $10^{-4}$ person$^{-1}$.year$^{-1}$. At that level, the impact of non-detect samples on the uncertainty of the risk estimate is limited. However, such monitoring is insufficient to verify a risk below the $10^{-4}$ level. When few *Cryptosporidium* oocysts are detected, the interpretation of the non-detect samples has a strong impact on the assessed risk level. Since such extensive monitoring is unable to verify risk levels below $10^{-4}$ and such monitoring is not feasible for all pathogens, QMRA will need to rely on source water monitoring and modelling of removal by treatment. The study has shown that removal of *Cryptosporidium* by conventional treatment systems can range from 1.8 to 5.2 log. More removal was found at higher *Cryptosporidium* concentrations in source water. This may be due to better design and operation of such plants to meet treatment standards or due to accumulation and release of *Cryptosporidium* in the treatment processes.

The use of a single average 'removal-credit' as applied in the Surface Water Treatment Rule can both underestimate or overestimate removal leading respectively to unnecessary actions or a false sense of safety. This underpins the need to perform site specific assessments of treatment performance. Since extensive microbial monitoring is not feasible for most sites, a treatment model needs to be developed that appropriately describes the treatment's ability to deal with peak events in source water, but also predicts the number of *Cryptosporidium* that break through treatment during nominal operation.

## ACKNOWLEDGEMENTS

The authors would like to thank the Water Companies in the UK that made their data available and to DWI for providing the statutory monitoring data.

## REFERENCES

Aboytes R., Di Giovanni G. D., Abrams F. A., Rheinecker C., McElroy W., Shaw N. and LeChevallier M. W. (2004). Detection of infectious *Cryptosporidium* in filtered drinking water. *J. Am. Water Works Assoc.*, **96**(9), 88–98.

Dugan N. R., Fox K. R., Owens J. H. and Miltner R. J. (2001). Controlling *Cryptosporidium* oocysts using conventional treatment. *Journal of the American Water Works Association*, **93**(12), 64–76.

DWI, Drinking Water Inspectorate (1999). *UK Water Supply Regulations (Amended) 1999*, www.dwi.gov.uk (accessed 1 March 2008).

DWI, Drinking Water Inspectorate (2006). personal communication.

Gibson III C. J., Haas C. N. and Rose J. B. (1999). Risk assessment of waterborne protozoa: current status and future trends. *Parasitology*, **117**, 205–212.

Haas C. N., Rose J. B. and Gerba C. P. (1999). *Quantitative Microbial Risk Assessment*. Wiley, New York, USA.

Hijnen W. A. M., Brouwer-Hanzens A. J., Charles K. J. and Medema G. J. (2005a). Transport of MS2 phage, *Escherichia coli*, *Clostridium perfringens*, *Cryptosporidium parvum*, and *Giardia intestinalis* in a gravel and a sandy soil. *Environmental Science and Technology*, **39**, 7860–7868.

Hijnen W. A. M., Beerendonk E. and Medema G. J. (2005b). Memo Including Rapid Sand Filtration, GAC Filtration and Direct Filtration to *Elimination of Microorganisms by Drinking Water Treatment Processes* (Not published). Kiwa Water Research, Nieuwegein, the Netherlands.

Hijnen W. A. M., Dullemont Y. J., Schijven J. F. and Medema G. J. (2006). Assessment of the Capacity of Slow Sand Filtration to Eliminate *Cryptosporidium* Oocysts. *Proceedings of Water Quality Technology Conference*, 5 to 9 November 2006, Denver, Colorado, AWWA.

Mons M. N, Van der Wielen J. M. L, Blokker E. J. M., Sinclair M. I., Hulshof K. F. A. M., Dangendorf F., Hunter P. R. and Medema G. J. (2007). Estimation of the consumption of cold tap water for microbiological risk assessment: an overview of studies and statistical analysis of data. *Journal of Water and Health*, **5**(S1), 151–170.

Nieminski E. C., Schaefer 3rd F. W. and Ongerth J. E. (1995). Comparison of two methods for detection of *Giardia* cysts and *Cryptosporidium* oocysts in water. *Applied and Environmental Microbiology*, **61**(5), 1714–1719.

Regli S, Rose J. B., Haas C. N. and Gerba C. P. (1991). Modeling the risk from *Giardia* and viruses in drinking water. *Journal of the American Water Works Association JAWWA5*, **83**(11), 76–84.

Teunis P. F. M., Medema G. J., Kruidenier L. and Havelaar A. H. (1997). Assessment of the risk of infection by *Cryptosporidium* or *Giardia* in drinking water from a surface water source. *Water Research*, **31**(6), 1333–1346.

Teunis P. F. M., Chappell C. L. and Ockhuysen P. C. (2002) *Cryptosporidium* dose-response studies: variation between isolates. *Risk Analysis*, **22**(1), 175–183.

USEPA (2006). LT2ESWTR, Long Term Second Enhanced Surface Water Treatment Rule.

WHO (2004). *Guidelines for Drinking Water Quality, (third edition)*. World Health Organization, Geneva, Switzerland.

Wohlsen T., Bates J., Gray B. and Katouli M. (2004). Evaluation of five membrane filtration methods for recovery of *Cryptosporidium* and *Giardia* isolates from water samples. *Applied and Environmental Microbiology*, **70**(4), 2318–2322.

# Chapter 4

# Inactivation of *Escherichia coli* by ozone under bench-scale plug flow and full-scale hydraulic conditions

*P.W.M.H. Smeets[a,d], A.W.C. van der Helm[b,c,d], Y.J. Dullemont[c], L.C. Rietveld[d], J.C. van Dijk[d] and G.J. Medema[a,d]*

[a]*KWR Watercycle research institute, PO Box 1072, 3430 BB Nieuwegein, the Netherlands*
[b]*DHV Water BV, PO Box 484, 3800 AL Amersfoort, the Netherlands*
[c]*Waternet, PO Box 8169, 1005 AD Amsterdam, the Netherlands*
[d]*Faculty of Civil Engineering, Delft University of Technology, PO Box 5048, 2600 GA Delft, the Netherlands*

## ABSTRACT

To determine the disinfection efficacy of ozonation, water companies can apply several disinfection calculation methods. The goal of this study was to evaluate the use of the T10 and CSTR method to extrapolate inactivation rates of ozone sensitive microorganisms observed in laboratory tests to full-scale ozonation in drinking water treatment. The inactivation efficacy of the ozonation at the Amsterdam water treatment works was assessed by determining *E. coli* concentrations in large volume samples before and after ozonation over a period of one year. The inactivation of dosed *E. coli* WR1 was tested in a bench-scale dissolved ozone plug-flow reactor (DOPFR) on the same feed water as the full-scale ozonation in which a concentrated ozone solution in Milli-Q® water was dosed. Applying the T10 method on the inactivation rates observed in the DOPFR strongly overestimated the inactivation capacity of the full-scale ozonation. The expected inactivation based on the CSTR method (LT2ESWTR) approached the observed inactivation at full-scale. Therefore, the CSTR method should be preferred to calculate inactivation of ozone sensitive organisms such as *E. coli*, viruses, *Giardia* and *Campylobacter* by full-scale ozonation.

 *Stochastic Modelling of Drinking Water Treatment in Quantitative Microbial Risk Assessment.* By Patrick Smeets. ISBN: 9781843393740. Published by IWA Publishing, London, UK.

## INTRODUCTION

Ozonation is widely used in drinking water treatment for disinfection, breakdown of micropollutants and odor and taste improvement. At the Amsterdam treatment works, ozonation forms an important barrier for bacteria, *Giardia* and viruses. To improve the disinfection capacity, increasing the ozone dose is being considered (Hijnen *et al.*, 2001). However, higher ozone doses also result in more bromate formation (Orlandini *et al.*, 1997). Inactivation of *Escherichia coli* (*E. coli*) by ozone has been studied by several researchers (Finch *et al.*, 1988; Zhou and Smith, 1994; Hunt and Mariñas, 1997 and Hijnen *et al.*, 2001). All have used different types of reactors (Continuous Stirred Tank Reactor CSTR, Plug Flow Reactor PFR, Batch Reactor) and different types of water. The results of these studies show a wide range of inactivation kinetics, even at (apparently) comparable conditions. For instance, the required *Ct* (ozone concentration times contact time) for 3.5 log inactivation in demand free buffer at 15–21°C was 0.001–0.05 mg min/l (see Table 4.1).

**Table 4.1** Reported ozone exposure (*Ct*) to reach 3.5–3.7 log inactivation of *E. coli*

| | | 2–5°C | 15–21°C |
|---|---|---|---|
| Farooq and Akhlaque, 1983 | lab waste water | | 0.3 |
| Finch *et al.*, 1988 | lab ozone demand free buffer | | 0.02 |
| Zhou and Smith, 1994 | lab ozone demand free buffer | 0.34 | 0.05 |
| Hunt and Mariñas, 1997 | lab ozone demand free buffer | 0.0025 | 0.001 |
| Hijnen *et al.*, 2001 | full-scale drinking water | 2.2 | 0.7[1] |

[1]*Only 2.6 log inactivation was found.*

Hijnen *et al.* (2001) compared the observed inactivation kinetics in full-scale ozone systems with inactivation kinetics determined under laboratory conditions. They concluded that ozonation in practice appears to be much less effective than could be expected from the lab-obtained inactivation kinetics. Suboptimal hydraulics of full-scale ozone systems are known to reduce the efficacy of ozonation in practice (Ducoste *et al.*, 2001).

In the Long Term Second Enhanced Surface Water Treatment Rule Toolbox Guidance Manual (LT2ESWTR, USEPA, 2003) two methods are used to calculate the inactivation efficacy of full-scale ozone contactors from inactivation rate constants determined in laboratory tests. The T10 method calculates the *Ct* by integrating the ozone profile in the contactor in time. The *Ct* is corrected to *Ct10* by applying a 'baffling factor' for short circuiting. The (extended) CSTR method assumes the contactor consists of a series of CSTRs. Inactivation is calculated from the measured ozone concentrations in each CSTR.

So far these methods are applied for *Cryptosporidium*, *Giardia* and viruses. The goal of this study was to evaluate the use of the T10 and CSTR method as models to extrapolate inactivation rates of ozone sensitive microorganisms such as *E. coli* observed in laboratory tests to full-scale ozonation in drinking water treatment.

## MATERIALS AND METHODS

### Bench-scale dissolved ozone plug flow reactor

Ozone was produced from medical grade air using a Fisher ozone generator model 503. A counter current glass bubble column with a diameter of 40 mm and a height of 1.0 m was operated at 9 l/h, 0.5 bar and 5°C to dissolve up to 50 mg/l ozone in Milli-Q® water. The ozone in Milli-Q® concentration was controlled by setting the ozone in gas concentration. Milli-Q® water with an electric conductivity (EC) of $5.5 \cdot 10^{-3}$ mS/m, and a total organic carbon (TOC) concentration of 5 μg/l was produced by a Milli-Q® Academic A10. TOC and EC were measured with an A10 TOC monitor (Millipore, USA). The ozone solution was fed through Teflon tubing to a venturi (test 1, 2) or a static mixer (test 3, 4, 5), which provided instant mixing with the test water. The Dissolved Ozone Plug Flow Reactor (DOPFR) (Figure 4.1) consisted of a 63.8 m polytertafluoroethylene (PTFE) tube of 8 mm internal diameter with 14 sampling points at different distances (Table 4.2). Test water was fed to the venturi by a circulation circuit with a pressure reducing valve to provide sufficient feed pressure. The length of the DOPFR was selected by opening a three-way valve thus directing the total flow through the sampling point.

## Full-scale installation

The full-scale installation (Figure 4.2) had a bubble column with a height of 5.0 m and a surface area of 5.7 $m^2$ followed by seven contact chambers with a total volume of 505 $m^3$. The average hydraulic residence time at a flow of 1500 $m^3$/h was 26.7 min. Ozone was produced by Trailigaz HRS150X ozone generators and was distributed in water with diffuser plates. Although the 6 sampling points were not placed in the ideal positions, ozone measured at these points was used to illustrate the magnitude of ozone concentrations at full-scale as compared to bench-scale. For *Ct* calculations the residence time was calculated from the contactor volume between sampling points and the flow through the contactor. When modeling inactivation, the ozone concentrations in the chambers were calculated and the position of the sampling points was not relevant.

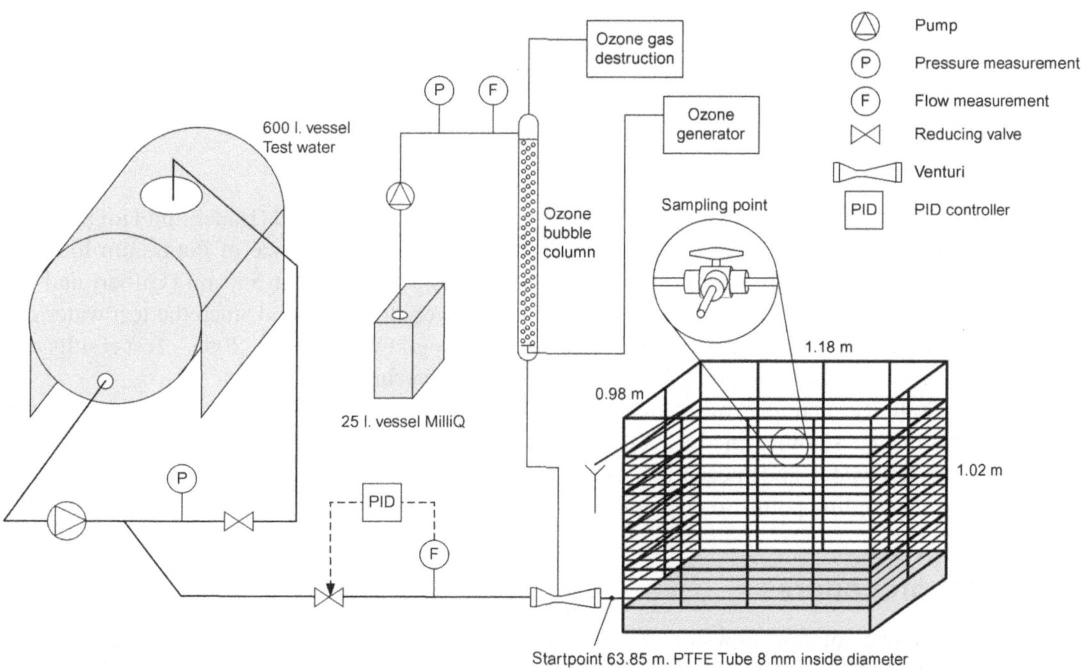


**Figure 4.1** Experimental setup of the DOPFR

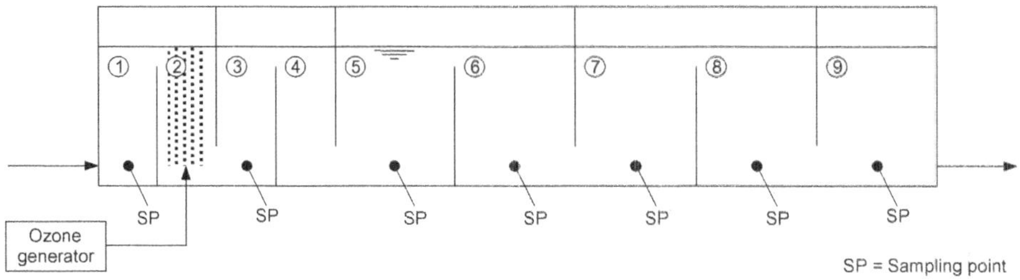


**Figure 4.2** The full-scale ozone contactor

## Microbial methods

*E. coli* WR1 strain was originally isolated from surface water and obtained from RIVM, Bilthoven, the Netherlands. An inoculum was prepared from a 24 h slant culture on agar using a pre-culturing technique with low substrate concentrations (Van der Kooij *et al.*, 1982). The inoculum was grown at 25°C in 1 l bottles containing autoclaved tap water supplemented with 1 mg acetate C/l, which reached maximum counts of approximately $7*10^5$ CFU/ml within one week. The bottles were stored at 6°C. A 1 l bottle was mixed with the test water (100–500 l) with an electrical mixer prior to experiments to reach approximately $3*10^3$ CFU/ml concentrations in the test water.

Microbiological samples were taken in 500 ml sterile bottles containing 4.2 ml 1% sterile sodiumthiosulphate to immediately quench any remaining ozone. Rapid mixing in the sample bottle was achieved by the powerful sample flow. Samples were analyzed by direct filtration and direct inoculation of the filter, or by dilution and direct inoculation of 0.1 ml on lauryl sulphate agar (Oxoid Basingstoke, England, nr. MM0615). Analysis in triplicate showed good reproducibility with a standard deviation of 13%. Large volume sampling at full-scale was conducted as previously described in Hijnen *et al.* (2000).

## Ozone analysis

Ozone in water for the pilot tests was determined according to the indigo method (Bader and Hoigné, 1982). In the full-scale installation ozone was analysed using DPD-sulphate in the presence of Potassium Iodide where the extinction of colour was determined with an UV/VIS-spectro-photometer at 550 nm (Gilbert and Hoigné, 1983). Both methods show cross-sensitivity to HOBr, which is likely to be formed since the test water contains bromide. This is relevant only at ozone concentrations below 0.1 mg/l (Von Gunten, 2005). Test results obtained below this concentration are presented in italic, since they are less accurate.

Conductivity, alkalinity, and pH were analyzed according to Dutch standard drinking water methods described in NEN. Ozone stability in the DOPFR was determined by using pure Milli-Q® water (without microorganisms) as test water and measuring ozone at the last sampling point. Ozone concentrations were measured at different sampling points to determine ozone decay in the DOPFR.

## Experimental procedures

The test water and ozone solution flow was 91 l/h and 9 l/h respectively. After switching sampling points the system was allowed to flush with 3 times the system volume before sampling. Test water was cooled or heated to the desired temperature. All flows were determined regularly by volume measurements. The experimental conditions are given in Table 4.2. Water quality parameters were stable, typical values are pH 8, alkalinity 205 mg/l, bromide 160 µg/l, EC 65 mS/m, dissolved organic carbon (DOC) 2.4 mg C/l, UV extinction at 254 nm (UV) 6.2 $m^{-1}$.

## Hydraulic model of the DOPFR

The residence time distribution (RTD) of the DOPFR was determined at each sampling point by step tracer experiments using a salt solution (120 g/l NaCl). The conductivity was simultaneously recorded every 0.1 s at two different points downstream from the tracer dosing point to determine the RTD between these points. A CSTR–in-series model was selected to describe the hydraulic characteristics of the DOPFR. The number of CSTRs and the hydraulic residence time were determined from RTD curves by parameter estimation through numerical integration of advection:

$$\left.\frac{dc}{dt}\right|_i^n = -v\frac{c_{i+1}^n - c_i^n}{\Delta x} \tag{1}$$

with:

$$n = \frac{L}{\Delta x} \tag{2}$$

$$t_h = \frac{L}{v} \tag{3}$$

$c$ is the tracer concentration (mg/l) ($c_0$ is influent concentration), $t$ is time (min), $n$ is the number of CSTRs, $v$ is the water velocity (m/min), $\Delta x$ is the length of one CSTR (m), $L$ is the length (m) and $t_h$ is the average hydraulic residence time (min) from point $i$ to $i + 1$.

The number of CSTRs and the hydraulic residence time were estimated through numerical integration of a set of $n$ finite-difference equations of advection (1) by fitting on measured residence time distribution curves. In this case the discretisation error, which is numerical dispersion, is assumed to be equal to the hydrodynamic dispersion (Peyret and Taylor, 1983). The relation between the dispersion coefficient $D$ ($m^2$/min), the number of CSTRs and the hydraulic residence time is:

$$D = v\frac{\Delta x}{2} = \frac{v \cdot L}{2 \cdot n} = \frac{L^2}{2 \cdot t_h \cdot n} \tag{4}$$

## Ozone profile calculations

A CSTR-in-series model (USEPA, 2003) was used to determine the slow ozone decay rate $k_s$ ($min^{-1}$) from the observed ozone concentrations in the DOPFR:

$$\frac{c_{O3,i}}{c_{O3,i-1}} = \left(\frac{1}{1 + k_s \bullet \frac{t_h}{n}}\right)^n \tag{5}$$

**Table 4.2** Test conditions and results of ozone and microbiological analysis; sampling point SP, tube length $L$ (m), cumulative volume $V$ (ml), number of CSTRs $n$, hydraulic residence time $t_h$, temperature $T$ (°C), ozone dose $c_{O3,d}$ (mg/l), analyzed ozone concentration $c_{O3,a}$ (mg/l) and $Ct$ (mg min/l)

| SP | $L$ m | $V$ ml | $n$ | $t_{10}/t_h$ | $t_h$ s | TEST 1 | | | TEST 2 | | | TEST 3 | | | TEST 4 | | |
|---|---|---|---|---|---|---|---|---|---|---|---|---|---|---|---|---|---|
| | | | | | | $T$ | 14 | | | 6 | | | 11 | | | 16 | |
| | | | | | | $c_{O3,d}$ | 0.85 | | | 0.41 | | | 0.43 | | | 0.70 | |
| | | | | | | pH | 7.97 | | | 8.13 | | | 8.05 | | | 7.99 | |
| | | | | | | $c_{O3,a}$ | $Ct$ | $DE$ | $c_{O3,a}$ | $Ct$ | $DE$ | $c_{O3,a}$ | $Ct$ | $DE$ | $c_{O3,a}$ | $Ct$ | $DE$ |
| 1 | 0.2 | 48 | 7 | 0.56 | 3 | | | | 0.18 | 0.01 | 1.2 | 0.18 | 0.00 | 2.0 | 0.21 | 0.01 | 3.11 |
| 2 | 0.9 | 106 | 23 | 0.74 | 3.6 | 0.31 | 0.02 | 4.27 | 0.14 | 0.01 | 2.3 | 0.12 | 0.01 | 3.2 | 0.19 | 0.01 | 3.83 |
| 3 | 1.8 | 179 | 27 | 0.76 | 6.4 | 0.21 | 0.02 | 5.00 | 0.11 | 0.02 | | 0.09 | 0.01 | | | | |
| 4 | 2.8 | 252 | 31 | 0.78 | 10.2 | 0.24 | 0.05 | 5.08 | 0.10 | 0.02 | 3.2 | 0.07 | 0.02 | 3.9 | | | |
| 5 | 3.7 | 325 | 30 | 0.77 | 13.7 | 0.22 | 0.06 | 5.30 | 0.08 | 0.03 | | 0.05 | 0.02 | | 0.13 | 0.03 | |
| 6 | 4.7 | 398 | 42 | 0.81 | 14.2 | 0.21 | 0.06 | 4.87 | 0.08 | 0.03 | 3.6 | 0.04 | 0.02 | 4.1 | | | |
| 7 | 7.3 | 555 | 50 | 0.82 | 22.2 | 0.09 | 0.04 | | 0.06 | 0.04 | | 0.02 | 0.02 | | | | |
| 8 | 9.9 | 711 | 106 | 0.88 | 25.9 | 0.07 | 0.04 | | 0.05 | 0.04 | 3.4 | 0.02 | 0.02 | | 0.07 | 0.04 | |
| 9 | 12.5 | 868 | 89 | 0.87 | 32.5 | 0.11 | 0.08 | | 0.05 | 0.05 | | 0.02 | 0.03 | | 0.06 | 0.05 | |
| 10 | 15.1 | 1023 | 135 | 0.89 | 40.1 | 0.11 | 0.11 | | 0.07 | 0.05 | | 0.03 | 0.03 | | | | |
| 11 | 23.3 | 1457 | 194 | 0.91 | 52.9 | 0.09 | 0.14 | | 0.05 | 0.06 | | 0.02 | 0.04 | | 0.05 | 0.09 | |
| 12 | 31.4 | 1892 | 230 | 0.92 | 68.1 | 0.07 | 0.18 | | 0.05 | 0.06 | | 0.02 | 0.04 | | | | |
| 13 | 47.5 | 2725 | 523 | 0.94 | 100.8 | 0.08 | 0.42 | | 0.04 | 0.06 | | 0.02 | 0.05 | | 0.05 | 0.25 | |
| 14 | 63.9 | 3576 | 725 | 0.95 | 130.1 | 0.09 | 0.84 | | 0.04 | 0.06 | | 0.01 | 0.06 | | 0.03 | 0.27 | |
| 15 | [1] | | | | 3600 | 0.00 | 1.14 | | | | | | | | 0.00 | 0.37 | |

[1] *Erlenmeyer vessel.*

$c_{O3,i}$ and $c_{O3,i-1}$ are the ozone concentrations at sampling point $i$ respectively a preceding sampling point $i-1$ (mg/l). The median of all $k_s$ values was used for further calculations.

The ozone dose $c_{O3,d}$ was calculated from the ozone solution concentration and the flow ratios. Instant ozone decay $c_{O3,instant}$ was calculated as:

$$c_{O3,instant} = c_{O3,d} - c_{O3,0} \quad (6)$$

$c_{O3,0}$ is the ozone concentration after instant ozone decay, extrapolated from $c_{O3}$ at sampling point 1. *Ct* (mg min/l) at point $i$ was calculated as the summed products of $c_{O3}$ and $t_h$ of each CSTR up to point $i$. *Ct10* (mg min/l) was calculated with the T10 method (USEPA, 2003) by applying a correction factor of $t_{10}/t_h$, where $t_{10}$ is the time at which 10% of the water has passed the contactor.

## Disinfection calculations

The inactivation of *E. coli* in an ideal batch or plug flow reactor can be described by a first order kinetic reaction (Oppenheimer *et al.*, 1999):

$$\frac{N_t}{N_0} = e^{-k_e Ct} \quad (7)$$

$N_0$ and $N_t$ are the number concentrations of organisms at 0 and t min respectively and $k_e$ is the inactivation rate constant (based on natural logarithm) in l/(mg min).

The CSTR equation (USEPA, 2003) is used to calculate inactivation in CSTR number $y$:

$$\frac{N_y}{N_y - 1} = \frac{1}{1 + k_e c_{O3,y} t_{h,y}} \quad (8)$$

$c_{O3,y}$ is the ozone concentration in CSTR number $y$ and $t_{h,y}$ is the hydraulic residence time in the CSTR. The total inactivation at a sampling point is calculated as the product of the inactivation in each of the CSTRs up to that sampling point:

$$\frac{N_i}{N_0} = \prod_0^n \frac{N_y}{N_{y-1}} \quad (9)$$

Disinfection efficacy is expressed as Decimal Elimination (*DE*):

$$DE = -\log\frac{N_i}{N_{i-1}} \quad (10)$$

Temperature has a strong effect on the inactivation rate of ozone. Several authors (Hunt and Mariñas, 1997 and Larson *et al.*, 2003) found that inactivation rates for ozonation follow Arrhenius' law:

$$k_e = A * e^{\left(\frac{-E_a}{RT_a}\right)} \quad (11)$$

$A$ is the frequency factor in l/(mg min), $E_a$ is the activation energy (J/mol), $R = 8.314$ J/(mol K) is the ideal gas constant and $T_a$ is the absolute temperature (K).

## RESULTS

### Hydraulic model of the DOPFR

Figure 4.3 shows the measured and calibrated conductivity of the RTD test for sampling point 4. The calibrated hydraulic residence times ($t_h$) of all sampling points in Table 4.2 are in compliance with the $t_h$ based on flow and volume. Sampling point 9 resulted in an unlikely low number of CSTRs (Table 4.2).

### Hydraulic model of the full-scale installation

The LT2ESWTR (USEPA, 2003) and the work of Do-Quang *et al.* (2000) suggest that each compartment can be modeled as a CSTR when simple baffles are used. The Amsterdam system has 7 compartments. Since the outlet of the last chamber is directly opposite to the inlet (Figure 4.2), it is not regarded to be a CSTR. A 6 CSTR model theoretically yields $t_{10}/t_h = 0.58$. Tracer tests at full-scale (Hofman *et al.*, 2005) resulted in $t_{10}/t_h = 0.55$–$0.65$, supporting the choice of a 6 CSTR model.

### Ozone decay

Ozone decay of the test water in Table 4.2 was characterized by an average instant ozone decay of 0.55 (0.4–0.6) mg/l within the first second after dosing, followed by first order ozone decay with average $k_s$ of 0.97 (0.9–1.0) $min^{-1}$.

Ozone stability tests showed that the ozone concentration at the last sampling point was 83% of the dose concentration. The ozone loss by auto decomposition, turbulence in the static mixer and/or by contact with the reactor surface was approximately 17%.

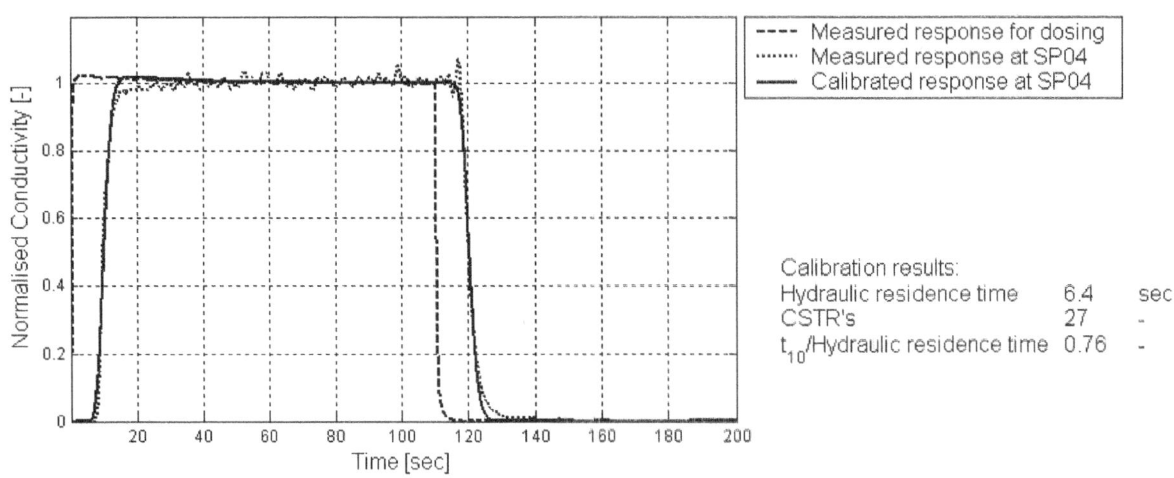


**Figure 4.3** Measured and calibrated step response for sampling point 4

**Table 4.3** Results of full-scale installation inactivation tests using large volume sampling for *E. coli* after ozonation; sample volume SV

| Date | Ct mg min/l | Ct10 mg min/l | T °C | *E. coli* in CFU/100 ml | SV l | *E. coli* out CFU/100 ml | DE |
|---|---|---|---|---|---|---|---|
| 5-02-03 | 1.77 | 1.06 | 5.9 | 3.7 | 25 | 0.0080 | 2.67 |
| 5-03-03 | 1.45 | 0.87 | 8.9 | 0.5 | 69 | <0.0014 | >2.54 |
| 2-04-03 | 1.67 | 1.00 | 10.4 | 0.1 | 53 | <0.0019 | >1.72 |
| 4-06-03 | 0.65 | 0.39 | 17.2 | 0.2 | 37 | <0.003 | >1.87 |
| 2-07-03 | 0.87 | 0.52 | 16.1 | 0.4 | 43 | <0.0047 | >2.24 |
| 6-08-03 | 0.75 | 0.45 | 18.1 | 1.9 | 33 | <0.0030 | >2.80 |
| 3-09-03 | 0.64 | 0.39 | 15.0 | 3.2 | 44 | <0.0023 | >3.15 |
| 1-10-03 | 1.25 | 0.75 | 14.3 | 0.3 | 43 | <0.0023 | >2.11 |
| 5-11-03 | 1.01 | 0.61 | 10.9 | 1.1 | 50 | 0.0120 | 2.36 |
| 1-12-03 | 1.28 | 0.77 | 9.8 | 0.55 | 31 | 0.0065 | 2.24 |
| 8-12-03 | 1.69 | 1.02 | 7.3 | 2.3 | 32 | 0.0125 | 2.26 |
| 15-12-03 | 1.29 | 0.78 | 8.5 | 2.3 | 31 | 0.0161 | 2.15 |
| 22-12-03 | 1.41 | 0.84 | 7.4 | 2.4 | 52 | 0.0115 | 2.49 |
| 29-12-03 | 1.38 | 0.83 | 7.8 | 0.8 | 52 | 0.0038 | 2.62 |
| 5-01-04 | 1.38 | 0.83 | 7.0 | 5.8 | 51 | 0.0059 | 2.99 |
| 12-01-04 | 1.32 | 0.79 | 8.5 | 0.4 | 53 | <0.0019 | >2.33 |
| 19-01-04 | 1.35 | 0.81 | 7.8 | 0.4 | 51 | <0.0020 | >2.31 |
| 26-01-04 | 0.84 | 0.50 | 7.2 | 0.8 | 51 | <0.0020 | >2.61 |
| 2-02-04 | 1.19 | 0.72 | 8.5 | 0.4 | 50 | 0.0040 | 2.00 |
| 9-02-04 | 0.98 | 0.59 | 7.4 | 0.5 | 50 | 0.0020 | >2.40 |

## *E. coli* inactivation in the DOPFR

Tests were performed to determine *E. coli* inactivation kinetics under conditions relevant for the Amsterdam water works. In natural water the ozone concentration quickly dropped below 0.1 mg/l. Still inactivation rate constants could be determined at reliable ozone concentrations (Table 4.2). At higher ozone doses, tailing was observed at a *DE* of 5 (Test 1 and 2 in Table 4.2).

## *E. coli* inactivation at full-scale

The inactivation results for the full-scale plant presented in Table 4.3 were calculated from twenty paired 1 l influent and 25 to 50 l effluent samples. *E. coli* was present in 10 of 20 samples after ozonation. At temperatures above 14°C no *E. coli* were detected in the treated water. The *DE* ranged between 2 and more than 3 at *Ct10* values between 0.4 and 1.0 (mg min)/l.

# DISCUSSION

## *E. coli* inactivation in literature

Figure 4.4 presents the Arrhenius plot for *E. coli* inactivation kinetics determined in the DOPFR and reported in literature. To provide the best model fit, data from Zhou and Smith (1994) was recalculated assuming a plug flow of 1 s caused by residence time in the syringe needle and non-ideal mixing conditions in the CSTR. It is clear that the temperature dependence of the inactivation is similar in all studies. The Hunt and Mariñas (1997) study shows significantly higher rate constants than the other studies. The studies used different *E. coli* strains and there were differences in culture conditions such as pH and temperature, which may explain the different inactivation rate constants. The DOPFR results agree well with the inactivation rate constants and temperature dependency found by Zhou and Smith (1994) and Finch *et al.* (1988). Therefore $E_a$ = 48,261 J/mol and log $A$ = 11.6 l/(mg min) are proposed for as conservative model values. The constants from Hunt and Mariñas (1997) ($E_a$ = 43,054 J/mol and log $A$ = 11.5 l/(mg min)) were used to determine the sensitivity of the model output to the inactivation rate constants.

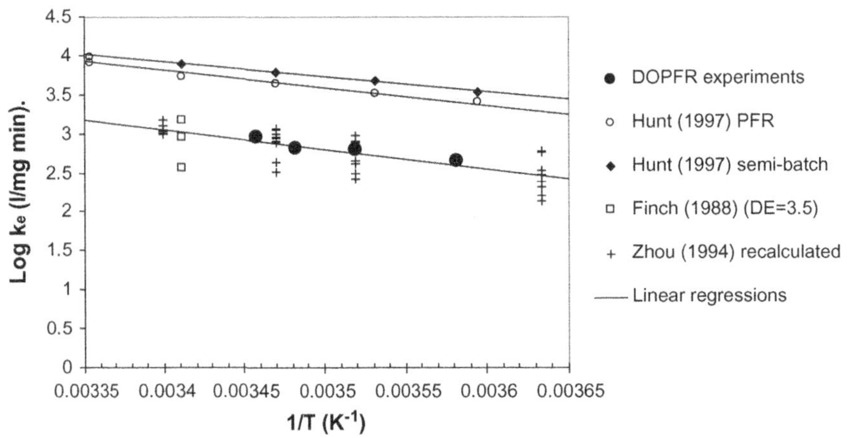


**Figure 4.4** *E. coli* inactivation rate constants from this study and literature

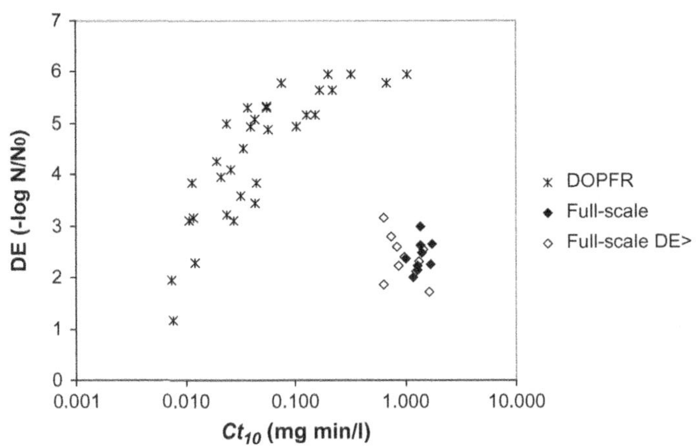


**Figure 4.5** *E. coli* inactivation in DOPFR and full-scale installation

## Comparing DOPFR and full-scale inactivation

The full-scale ozone installation consisted of five parallel lines in all of which ozone was measured. From the ozone profile the *Ct10* was calculated for each line with a $t_{10}/t_h$ of 0.6. Microbiological samples were taken from the combined flow of all five contactors before and after ozonation. Figure 4.5 shows the observed inactivation in the DOPFR and the full-scale ozonation. Negative post ozonation *E. coli* samples are presented as '*DE* larger than' (*DE*>) values, calculated by assuming 1 colony counted in the largest (negative) sample. The reported *Ct10* in Figure 4.5 (on log scale) is the average *Ct10* of the five ozone lines. Although ozone at full-scale was not measured at ideal positions, the inactivation at similar *Ct* values was clearly much higher in the DOPFR than at full-scale.

## T10 and CSTR calculations for *E. coli* inactivation in full-scale plant

Expected *E. coli* inactivation at the Amsterdam treatment plant was calculated with the T10 and the CSTR method. Table 4.4 shows that the T10 method strongly overestimates disinfection efficacy. CSTR model calculations with the constants form Hunt and Mariñas (1997) still strongly overestimates inactivation, although it is better than the T10 method. The CSTR calculations with the proposed kinetics accounted for most of the difference between bench-scale and full-scale observations. Still there is a significant difference between the calculated inactivation with the CSTR model and the observed inactivation. Possible explanations are the heterogeneous susceptibility of environmental *E. coli* (Hom, 1972), age distribution in a population (Chick, 1908) or protection by encapsulation in aggregates of microorganisms or particles (Hijnen *et al.*, 2004). Haas and Kaymak (2003) suggest that high initial microbial densities in disinfection experiments result in higher inactivation rate constants. Measurements over the cross-section of a pilot-scale conventional contactor showed that water is poorly mixed in and after the bubble column. This results in zones with ozone concentrations of <10% to 150% of average concentrations (Van der Veer *et al.*, 2005). Since the water is poorly mixed in the contact chambers, zones with low ozone concentrations will provide little inactivation. This introduces another hydraulic shortcoming of conventional contactors.

**Table 4.4** Observed and modeled *DE* (log inactivation) of *E. coli* in the full-scale installation either applying the CSTR or the T10 method; $c_{O3,d}$ 0.8 mg/l, $c_{O3,instant}$ 0.5 mg/l, $k_s$ 1 min$^{-1}$

| | Temperature | | | | | |
|---|---|---|---|---|---|---|
| | 0 | 5 | 10 | 15 | 20 | 25 |
| Observed | – | 2.6 | 2.3 | >3.1 | >2.8 | – |
| T10 | 19 | 27 | 40 | 56 | 79 | 111 |
| CSTR (Hunt and Mariñas[1]) | 8.0 | 8.8 | 9.5 | 10.3 | 11.0 | 11.7 |
| CSTR (proposed) | 4.3 | 4.9 | 5.4 | 6.2 | 6.9 | 7.6 |

[1] *Inactivation kinetics found by Hunt and Mariñas (1997) were applied.*

## Significance for water treatment

To illustrate the significance of these findings for water treatment, the expected inactivation of *Cryptosporidium*, *Giardia*, viruses and *E. coli* was calculated for the Amsterdam ozone system. Ozone inactivation kinetics in Table 4.5 were used. Assuming an ozone dose of 2 mg/l, instant ozone decay of 0.5 mg/l and slow ozone decay of 1 min$^{-1}$ the ozone profile was calculated. This would result in a *Ct10* of 0.9 for the Amsterdam contactor. According to the T10 method log inactivation of *Cryptosporidium*, *Giardia*, viruses and *E. coli* would be 0.09, 1.9, 3.9 and 195 respectively. However, when the CSTR method was applied, log inactivation estimates were 0.14, 1.4, 2.1 and 8.7 respectively. Figure 4.6 shows how the inactivation of the different microorganisms depends on the number of CSTRs. The CSTR model calculates less inactivation of *Giardia* and viruses than the T10 method for systems of less than 20 or 50 CSTRs respectively. The T10 method should not be used for full-scale plants since they are generally characterized by 3 to 7 CSTRs in series (Do Quang *et al.*, 2000).

**Table 4.5** Inactivation rates of modelled organisms

| Organism | $k_e$ 10°C (l/mg min) | Reference |
|---|---|---|
| *E. coli* | 499 | DOPFR/Zhou and Smith (1994) |
| *Cryptosporidium* | 0.24 | USEPA (2003) |
| *Giardia* | 4.9 | AWWA (1991) |
| viruses | 10.0 | AWWA (1991) |

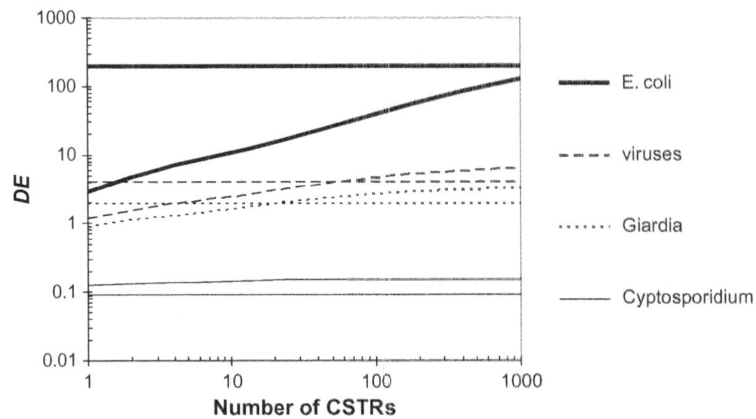


**Figure 4.6** Predicted inactivation of different organisms with CSTR method (curved lines), and T10 inactivation credits (horizontal lines); T 10°C, $c_{O3,d}$ 2 mg/l, $k_s$ 1.0 min$^{-1}$, $c_{O3,instant}$ 0.5 mg/l, $t_h$ 15 min, $t_{10}/t_h$ 0.6

Amsterdam water works considered increasing the ozone dose to improve disinfection of *E. coli*. Therefore the same calculations were performed but with a fixed number of 6 CSTRs and different ozone doses. Ozone concentrations were calculated from the ozone dose and the ozone decay rate. Figure 4.7 illustrates how the effect of increasing the ozone dose on inactivation of *E. coli*, viruses and *Giardia* is limited by the hydraulics of the system. For a resistant organism such as *Cryptosporidium*, the hydraulics have little effect on the inactivation kinetics. Since little inactivation takes place in the bulk of the water, small volumes with no inactivation have little effect on the average concentration after ozonation.

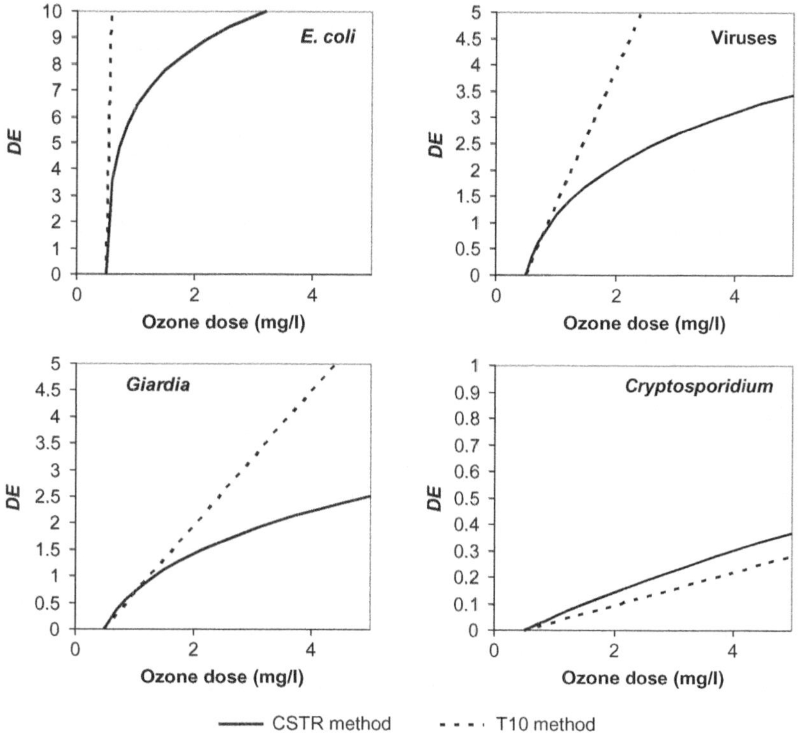


**Figure 4.7** CSTR and T10 inactivation credits versus ozone dose in a conventional full-scale (6 CSTR) system; $T$ 10°C, $k_s$ 1.0 min$^{-1}$, $c_{O3,instant}$ 0.5 mg/l, $t_h$ 15 min, $t_{10}/t_h$ 0.6

## CONCLUSIONS

Dosing ozone in the form of a concentrated solution in a DOPFR provides higher *E. coli* inactivation than in a conventional ozone contactor. Hydraulic conditions in conventional ozone contactors show a strong deviation from plug flow, resulting in far less inactivation of ozone sensitive organisms than expected based on LT2ESWTR T10 calculations. The application of the CSTR method improves prediction of inactivation of microorganisms by full-scale ozonation and is therefore recommended. Still inactivation is overestimated, possibly due to poor mixing of ozone over the water volume resulting in streamlines with little or no ozone and thus little or no disinfection. Increased resistance of sub-populations or particle protection can also reduce inactivation. Improving plug flow in a conventional ozone contactor should be considered in stead of increasing ozone dose to improve the disinfection capacity of a conventional ozone contactor. This is now studied at the Amsterdam Water Supply.

## REFERENCES

AWWA (American Water Works Association) (1991). *Guidance Manual for Compliance with the Filtration and Disinfection Requirements for Public Water Systems Using Surface Water Sources*. American Water Works Association, Denver, CO.

Bader H. and Hoigne L. (1982). Determination of ozone in water by the indgo method, a submitted standard method. *Ozone: Science and Engineering*, **4**, 169–176.

Chick H. (1908). An investigation of the laws of disinfection. *Journal of Hygiene*, Cambridge, **8**, 92–158.

Do-Quang Z., Ramirez C. C. and Roustan M. (2000). Influence of geometrical characteristics and operating conditions on the effectiveness of ozone contacting in fine-bubbles conventional diffusion reactors. *Ozone: Science and Engineering*, **22**, 369–378.

Ducoste J., Carlson K. and Bellamy W. (2001). The integrated disinfection design framework approach to reactor hydraulics characterization. *Aqua – Journal of Water Supply: Research and Technology*, **50**(4), 245–261.

Farooq S. and Akhlaque S. (1983). Comparative response of mixed cultures of bacteria and virus to ozonation. *Water Research*, **17**(7), 809–812.

Finch G. R., Smith D. W. and Stiles M. E. (1988). Dose-respons of *Escherichia coli* in ozone demand-free phosphate buffer. *Water Research*, **22**, 1563–1570.

Gallard H., von Gunten U. and Kaiser H. P. (2003). Prediction of the disinfection and oxidation efficiency of full-scale ozone reactors. *Journal of Water Supply: Research Technology-AQUA*, **52**(4), 277–290.

Gilbert E. and Hoigné J. (1983). Messung von ozon in wasserwerken: Vergleich der DPD- und indigo-methode. *GWF-Wasser/Abwasser*, **124**(11), 527–531.

Haas C. N. and Kaymak B. (2003). Effect of initial microbial density on inactivation of *Giardia muris* by ozone. *Water Research*, **37**, 2980–2988.

Hijnen W. A. M, Veenendaal D., van der Speld W. M. H., Visser A., Hoogenboezem W. and van der Kooij D. (2000). Enumeration of faecal indicator bacteria in large water volumes using on site membrane filtration to assess water treatment efficiency. *Water Research*, **34**(5), 1659–1665.

Hijnen W. A. M., Bosklopper Th. G. J., Hofman J. A. M. H., Bosch A. D. and Medema G. J. (2001). Improvement of the disinfection efficiency of the full-scale ozonation of the River-lake waterworks of Amsterdam Water Supply. In *Proceedings of the International Ozone Association Congress*, September 2001, London.

Hijnen W. A. M., Baars E., van der Veer A. J., Bosklopper Th. G. J., Meijers R. T. and Medema G. J. (2004). Influence of DOC on the inactivation efficiency of ozonation assessed with *Clostridium perfringens* and a lab-scale continuous flow system. *Ozone: Science and Engineering*, **26**(5), 465–473.

Hofman J. A. M. H., Smeets P. W. M. H., Van der Veer A. J., Gijsbertsen-Abrahamse A. J., Wind A. P. M. and van Dijk J. C. (2005). Improvement of ozone disinfection: application of integrated flow modeling and full-scale validation. *WQTC conference 6–10 November 2005*, Quebec, Canada,

Hom L. W. (1972). Kinetics of chlorine disinfection in an ecosystem. *Journal of Sanitary Engineering Division, ASCE*, **98**, 183–193.

Hunt N. K. and Mariñas B. J. (1997). Kinetics of *Escherichia coli* inactivation with ozone. *Water Research*, **31**, 1355–1362.

Larson M. A. and Mariñas B. J. (2003). Inactivation of *Bacillus subtilis* spores with ozone and monochloramine. *Water Research*, **37**, 833–844.

NEN, Dutch Normalization Institute, Vlinderweg 6, 2623 AX, Delft, the Netherlands, www.nen.nl (accessed 1 March 2008).

Oppenheimer J. A., Aieta E. M., Trussel R. R., Jacangelo J. G. and Najm I. N. (1999). *Evaluation of Cryptosporidium Inactivation in natural waters.* AWWA Research Foundation Denver, Colorado, U.S.A.

Orlandini E., Kruithof J. C., van der Hoek J. P., Siebel M. A. and Schippers J. C. (1997). Impact of ozonation on disinfection and formation of biodegradable organic matter and bromate. *Journal of Water Supply: Research Technology-AQUA*, **46**, 20–30.

Peyret R. and Taylor T. D. (1983). *Computational Methods for Fluid Flow.* Springer-Verlag, New York.

USEPA (2003). *LT2ESWTR Long Term Second Enhanced Surface Water Treatment Rule and Draft Toolbox Guidance Manual USEPA*, Washington DC, USA.

Van der Kooij D., Visser A. and Hijnen W. A. M. (1982). Determining the concentration of easily assimilable organic carbon in drinking water. *Journal of Water Supply: Research Technology-AQUA*, 540–545.

Van der Veer A. J., Schotsman R. M., Martijn B., Schurer R., van Dijk J. C. (2005). Comparative small technical Scale Study on Ozone Disc Diffusers and Ejector System. *17th IOA conference 22–26 August 2005*, Strasbourg, France.

Von Gunten U. (2005). Eawag, Ueberlandstrasse 133 P. O. Box 611 8600 Duebendorf, Switzerland. Personal communication.

Zhou H. and Smith D. W. (1994). Kinetics of ozone disinfection in completely mixed system. *Journal of Environmental Engineering*, **120**(4), 841–858.

# Chapter 5

# Improved methods for modelling drinking water treatment in quantitative microbial risk assessment; a case study of *Campylobacter* reduction by filtration and ozonation

*P.W.M.H. Smeets[a,c], Y.J. Dullemont[b], P.H.A.J.M. van Gelder[c], J.C. van Dijk[c] and G.J. Medema[a,c]*

*[a]KWR Watercycle research institute, PO Box 1072, 3430 BB Nieuwegein, the Netherlands*
*[b]Waternet, PO Box 8169, 1005 AD Amsterdam, the Netherlands*
*[c]Faculty of Civil Engineering, Delft University of Technology, PO Box 5048, 2600 GA Delft, the Netherlands*

## ABSTRACT

Quantitative microbial risk assessment (QMRA) is increasingly applied to estimate drinking water safety. In QMRA the risk of infection is calculated from pathogen concentrations in drinking water, water consumption and dose-response relations. Pathogen concentrations in drinking water are generally below detection limits and monitoring the drinking water for pathogens provides little information for QMRA. Therefore pathogen concentrations are monitored in the raw water and reduction of pathogens by treatment is modelled stochastically with Monte Carlo simulations. The method was tested in a case study with *Campylobacter* monitoring data of rapid sand filtration and ozonation processes. This study showed that the currently applied method did not predict the monitoring data used for validation. Consequently the risk of infection was overestimated by one order of magnitude. An improved method for model validation was developed. It combines non-parametric bootstrapping with statistical extrapolation to rare events. Evaluation of the treatment model was improved by presenting monitoring data and modelling results in complementary cumulative distribution function (CCDF) graphs, which focus on the occurrence of rare events. Apart from calculating the yearly average risk of infection, the model results were presented in frequency number (FN) curves. This allowed for evaluation of both the distribution of risk and the uncertainty associated with the assessment.

 *Stochastic Modelling of Drinking Water Treatment in Quantitative Microbial Risk Assessment.* By Patrick Smeets. ISBN: 9781843393740. Published by IWA Publishing, London, UK.

## INTRODUCTION

Monitoring the absence of indicator organisms in drinking water has been the main approach to safeguard drinking water quality since the beginning of the 20th century (Greenwood and Yule, 1917). However, drinking water outbreaks of infectious disease have shown that absence of indicator organisms in drinking water does not imply that there is no risk of infection (Hrudey and Hrudey, 2004). Since 1980 Quantitative Microbial Risk Assessment (QMRA) has been applied to quantify the microbial safety of drinking water (Haas, 1983; Gerba *et al.*, 1988; Regli *et al.*, 1991; Rose *et al.*, 1991; Teunis *et al.*, 1994; ILSI, 1996; Gibson III *et al.*, 1999 and Payment *et al.*, 2000). Risk of infection is calculated from the chance of ingesting pathogens (exposure or dose) and the chance of developing an infection from this exposure (dose-response relation) (Haas *et al.*, 1999). Pathogen concentrations in drinking water are generally below detection limits (Regli *et al.*, 1991). QMRA studies have therefore monitored pathogen concentrations in the raw water and modelled removal or inactivation by treatment to estimate concentrations in the drinking water (Teunis *et al.*, 1997; Haas *et al.*, 1998; Teunis and Havelaar, 1999 and Westrell *et al.*, 2003).

In these QMRA studies, variability of each element was described by a Probability Density Function (PDF). Treatment was then stochastically modelled by Monte Carlo simulation. Determining the PDF for each element using the available data is a crucial step in such an assessment. PDF parameters have been estimated from pilot study results or literature. However, since raw water concentration and treatment efficacy vary in time and are specific for each drinking water production location, site specific information is preferred (Teunis *et al.*, 1997; Nichols, 2003 and Smeets *et al.*, 2007). Monitoring pathogens or indicator organisms in raw water and after treatment steps provides such information. QMRA studies have fitted statistical distributions to such data to determine the PDF. Drinking water risk assessments have mainly used the lognormal, gamma and negative binomial distributions (Teunis *et al.*, 1997 and Haas *et al.*, 1999). Other fields of risk assessment commonly use the Weibull distribution (Van Gelder, 1999). The impact of choice of distribution on the result of the risk assessment has not been studied extensively (Haas *et al.*, 1997). Preliminary trials to the current study indicated that the choice of PDF could dominate the QMRA outcome. Therefore this study focussed on non-parametric techniques for QMRA that do not require an a priori choice of PDF (Van Gelder, 1999).

Previous studies (Teunis *et al.*, 2004) have shown that extreme events can dominate the average health risk. Historical data on source water concentrations and treatment efficacy can be used to predict normal rare events. These events are caused by the extremes of normal variations in the system such as flow changes, rainfall events, seasonal variations and treatment variations. Observed normal variations are extrapolated to these extreme events by a PDF. Therefore PDF should fit the extremes (tail) of observed variation, in this case monitoring data, since it is used to predict rare events of high concentrations or poor treatment. However, current methods of PDF estimation focus on the distribution type and parameters that best describe the bulk of the data, such as the Kolmogorov-Smirnov test (Haas *et al.*, 1999) or likelihood ratio (Teunis *et al.*, 1997). This study adopted the use of Complementary Cumulative Distribution Functions (CCDF) graphs (Van Gelder, 1999) to visually evaluate the fit of PDF to the tail of the data.

In current QMRA practice the treatment efficacy PDF is validated based on fractions resulting from microbial counts before and after treatment in samples taken on the same date. However, preliminary studies by the first author showed that the predicted concentrations were not in line with the monitored concentrations. Therefore improved methods for model validation were developed in this study.

Much focus in QMRA studies has been on accounting for sampling variability due to (over-) dispersion, variable recovery, pathogen viability and infectivity (Teunis *et al.*, 1997; Haas *et al.*, 1999 and Teunis and Havelaar, 1999). The uncertainty that is introduced by Most Probable Number (MPN) type data has not been well studied. Haas *et al.* (1999) treated MPN data similar to count data. Although an 85% correction factor was applied to account for bias in the reported MPN it did not include the uncertainty of the MPN in the risk assessment. Since the case study included MPN type data, a method to include MPN uncertainty was developed.

The outcome of QMRA studies is generally presented as a PDF or histogram of risk of infection (Westrell *et al.*, 2003). No distinction between variability and uncertainty was thus made. Other fields of risk assessment such as flooding, traffic or industrial accidents, present societal risk of major accidents in a FN-curve (Van Gelder, 1999) plotting the number of casualties (N) versus the frequency of occurrence (F). This method seems appropriate for assessment of risk of infection through drinking water, since it is a societal risk. The FN curve allows differentiating between low incidental risk (1 infection per 10,000 people each day of the year) and an outbreak (e.g. 365 infections per 10,000 people on one day in a year). Although the yearly average risk is identical in both situations, the outbreak is considered less acceptable than the incidental risk. Therefore the FN curve provides better decision support for risk managers and inspectors than a distribution of the yearly average risk.

Methods for large volume sampling, up to 1,000 L, have become available in recent years (Hijnen *et al.*, 2000 and Smeets *et al.*, 2007). Since resources are limited, water utilities need to carefully plan their sampling strategy, which includes finding a balance between a limited number of large volume samples and a larger number of regular volume samples. This study differentiated what concentrations are most relevant for the yearly average risk of infection in order to support such decisions.

The goal of this study was to develop improved methods for modelling drinking water treatment in quantitative microbial risk assessment of drinking water and to apply these methods in a case study. The following methods were adopted from other fields of risk assessment or newly developed:

- Non-parametric bootstrap method for data uncertainty analysis;
- Including MPN uncertainty in the non-parametric bootstrap method;
- Implementation of CCDF graphs for data presentation;
- Verification of validation method (model outcome matches the validation data);
- PDF fitting with focus on tails of data;
- Determination of relative risk related to concentrations;
- Implementation of FN curves for risk presentation.

The paper first describes the methods and the case study and compares different methods of data presentation. Then the non-parametric bootstrap method was applied to determine data uncertainty, including MPN uncertainty. Next the currently applied method to validate pathogen reduction by treatment for Monte Carlo simulation was compared to improved methods. The validated non-parametric treatment model was applied to predict pathogen concentrations after treatment. By comparing the predicted concentrations to the monitored concentrations, the accuracy of the current and improved methods was compared. Next parametric distributions were fitted to the validated model to extrapolate to rare events of high raw water pathogen concentrations or poor reduction by treatment. The predicted concentrations of the parametric treatment model were also compared to the monitored concentrations to verify the accuracy of the model. Risk of infection was then calculated from the concentrations predicted both with the currently applied method and the improved method of treatment model validation. Risk of infection was determined for each concentration to assess the relative impact on the risk, which provided guidance for monitoring. Finally the risk assessed with current and improved methods was compared in a FN curve.

## METHODS

### Case description

*Campylobacter* monitoring data collected at the water treatment plant (WTP) Leiduin of Waternet (water cycle company for Amsterdam and surrounding areas) was used for the case study. The source for the WTP is water from the river Rhine which is pre-treated and infiltrated in the dunes. The water is abstracted in open canals and collected in an open reservoir before post-treatment. The pre-treatment and the soil passage remove most pathogens from the Rhine water; however the water is re-contaminated by birds and wildlife in the open canals and reservoir directly before the treatment plant intake. The water in the reservoir was referred to as raw water in the case study. The reservoir is situated in a protected dune area, therefore contamination of the reservoir through waste water and agricultural run-off is unlikely. Water fowl like ducks, geese, gulls and swans in the reservoir and the abstraction canals are the most likely sources of *Campylobacter*. Wildlife like deer, rabbits and rodents and possibly some pets (cats, dogs) in the dunes can also contribute to the contamination with *Campylobacter*. Contamination can take place either by entering the water (waterfowl, rats, dogs) or by shedding faeces on the shore, which is then washed into the reservoir by rain during run-off. The reservoir is refreshed daily. Since the contamination takes place only in a small proportion of the water (the water surface where the ducks are swimming and the shores) and the reservoir is not mixed, the water quality at the intake sampling point is likely to vary significantly. The raw water is treated by rapid sand filtration, ozonation, softening, biological activated carbon filtration and slow sand filtration. Rapid sand filtration, ozonation and slow sand filtration are considered the main microbial barriers at the WTP (Figure 5.1). The risk of infection was calculated for consumption of ozonated water. Since some *Campylobacter* were detected in ozonated water this dataset was more appropriate to demonstrate the improved methods than the drinking water dataset in which no *Campylobacter* was detected.

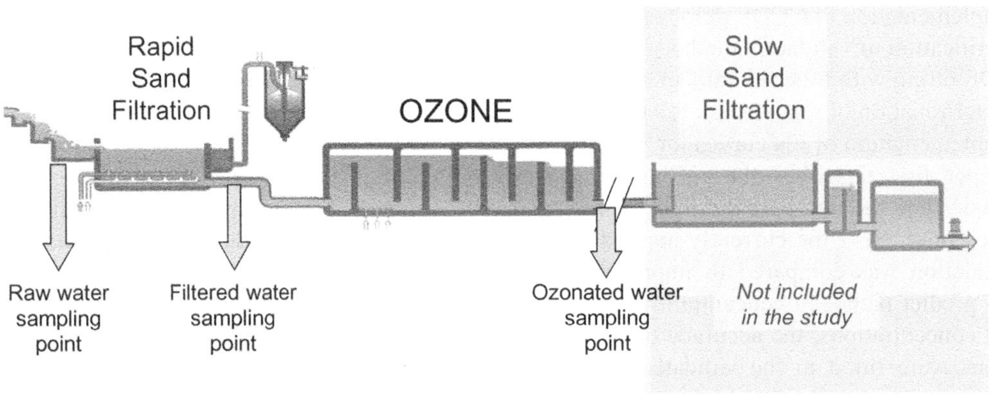


**Figure 5.1** Microbial barriers and location of sampling points at WTP Leiduin

## Microbial analysis

*Campylobacter* samples were analyzed by direct filtration and direct inoculation of the filter in tubes with Preston *Campylobacter* selective enrichment broth. Positive results were confirmed by microscopic examination of a hanging drop for the presence of *Campylobacter*. *Campylobacter* was quantified by the most probable number method (MPN) in three parallel tubes for three filtered sample volumes using decimal dilutions. The collected monitoring data consisted of the MPN-arrays for each sample (e.g. 3-2-1 indicated three positive tubes in the largest volume, two positive tubes in the middle volume and one positive tube in the smallest volume). Reported MPN's were taken from MPN tables by De Man (1975).

## Non-parametric MPN bootstrapping

The bootstrap method is a fairly easy tool to numerically calculate the uncertainty of a dataset of measurements, by repeatedly drawing results randomly from the dataset. The confidence interval for the monitored *Campylobacter* concentrations was determined by adapting a standard non-parametric bootstrapping procedure (Van Gelder, 1999) to include MPN method uncertainty (De Man, 1975). A result was randomly drawn from the $m$ monitoring results (with replacement, Equation 1) and for this result a random concentration was drawn according to the MPN likelihood distribution for the result (Equation 2). Thus a bootstrap sample of *Campylobacter* concentrations in the monitored water was produced.

$$X_{ij}^{*} = X_{[m.p]} \qquad i = 1,\ldots,n \quad j = 1,\ldots,k \tag{1}$$

$$\begin{aligned} q(C^{*}_{ij}|X^{*}_{ij}) = {} & (1 - e^{-v_1 C_{ij}})^{P_1}(1 - e^{-v_1 C_{ij}})^{3-P_1}(1 - e^{-v_2 C_{ij}})^{P_2}\ldots \\ & (1 - e^{-v_2 C_{ij}})^{3-P_2}(1 - e^{-v_3 C_{ij}})^{P_3}(1 - e^{-v_3 C_{ij}})^{3-P_3} \end{aligned} \tag{2}$$

Where $X$ is the dataset of monitored MPN results, $X^*$ is the bootstrap dataset of MPN results, $i$ indicates the $i$th MPN result, $j$ indicates the $j$th bootstrap sample, $n$ is the number of draws in a bootstrap sample, $k$ is the total number of bootstrap samples in a bootstrap dataset, $p$ and $q$ are uniform random variables, *[m.p]* is the integer ceil function (round up of *m.p*), $C^*$ is the bootstrap sample of *Campylobacter* concentrations (organisms/L), $q$ is the likelihood of concentration $C^*$ given result $X^*$, $v_1$, $v_2$, $v_3$ are the three volumes (or dilutions) used in the MPN method and $P_1$, $P_2$, $P_3$ are the number of positives at the respective volumes. Equation 2 was solved numerically to determine $C^*$ at a given $q$ and $X^*$. By producing $k$ bootstrap samples of size $n$ with $n = m$ the $C^*$ resembled the likelihood of *Campylobacter* concentrations given the presence/absence results. From this the 95% confidence interval (CI) of the concentration was determined for each proportion of the water. Stable results were achieved with acceptable calculation times for $k = 10{,}000$.

In some ozonated water samples no *Campylobacter* were detected (0-0-0 result). Consequently the MPN likelihood $q$ in Equation 2 approaches 1 as $C$ approaches 0, so no lower limit of the likely concentration can be determined. As a practical approach, the non detects were adapted before bootstrapping by doubling the sample volume and assuming one positive in the largest MPN volume (1-0-0 result). This is similar to setting non-detect samples of count data to 'half the detection limit', which is a conservative approach in risk assessment. Similarly, raw water samples that were all positive (3-3-3 result) were adapted to one negative in the smallest MPN volume (3-3-2 result) with half the sample volume to provide an upper limit of likely concentration. Although this is a simplified approach for these 'larger than' values, it proved to be efficient to demonstrate the methods in this study. Preferably these issues are prevented during monitoring by using sufficient sample dilutions. The bootstrap samples of raw, filtered and ozonated water were used for the assessment of pathogen reduction by treatment, the assessment of the raw water PDF and model verification.

## Non-parametric validation of treatment efficacy

Treatment efficacy $\pi$ is the fraction of organisms that pass a treatment step. The observed treatment efficacy was calculated from the bootstrap datasets of monitoring data as:

$$\pi_{ij}^{*} = \frac{C^{*}_{out\ [n.p1][k.p2]}}{C^{*}_{in\ [n.p1][k.p3]}} \qquad i = 1,\ldots,n \quad j = 1,\ldots,k \tag{3}$$

Where *p1*, *p2* and *p3* are uniform random variables and *[n.p1]* is the integer ceil function. Several methods can be used to select values from the bootstrap samples $C_{in}^*$ and $C_{out}^*$ that are 'paired' in Equation 3. The effects of using the 'random', 'date' or 'rank' method was studied. The bootstrap samples $C_{in}^*$ and $C_{out}^*$ require different preparations for these methods.

The random method assumes no correlation by date or rank. The bootstrap samples $C_{in}^*$ and $C_{out}^*$ did not undergo any adaptation, so samples before and after treatment were paired randomly (since samples $X^*$ were selected randomly in Equation 1).

Pairing by date has been widely applied in QMRA (Teunis *et al.*, 1997; Teunis and Havelaar, 1999 and Teunis *et al.*, 1999) and can be considered the current 'state of the art'. Influent and effluent samples taken on the same day are compared and $\pi$ is calculated for each pair. This assumes that samples before and after treatment are correlated in time. To enable pairing by date, the monitoring datasets $X_{in}$ and $X_{out}$ were prepared so that they only included results taken on the same day in date order. Equation 1 was adapted so that samples were drawn in order and without replacement (*[m.p]* was replaced by *i* so all bootstrap samples included every result once). Effectively the date-bootstrap procedure produced a bootstrap dataset that only included MPN uncertainty.

Pairing by rank has only been reported once (Teunis *et al.*, 1999) and was referred to as 'unpaired counts', but its application was not explored further. Pairing by rank assumes complete correlation between the influent and effluent concentrations (lowest influent concentrations correlate to lowest effluent concentrations etc.). To enable pairing by rank, the bootstrap samples $C_{in}^*$ and $C_{out}^*$ were sorted by concentration before determining $\pi$.

Using Equations 1, 2 and 3 $\pi^*_{filt}$ was determined from $C^*_{raw}$ and $C^*_{filt}$, and $\pi^*_{O3}$ was determined from $C^*_{filt}$ and $C^*_{O3}$. Thus $\pi^*$ resembled the likelihood of actual *Campylobacter* reduction by removal and inactivation respectively. From this the 95% confidence interval (CI) of the reduction was determined for each proportion of the water for presentation in graphs. The study used the total bootstraps in calculations, not the 95% CI.

## Parametric extrapolation of bootstrap samples

Parametric distributions were fitted to the $k$ bootstrap samples of $n$ raw water $C^*_{raw}$, removal $\pi^*_{filt}$ and inactivation $\pi^*_{O3}$ values respectively using the fit functions in Matlab® for several distribution types. This resulted in $k$ parameter pairs for each distribution type. Gamma, lognormal and Weibull distributions were fitted to $C^*_{in}$, $\pi^*_{filt}$ and $\pi^*_{O3}$. The beta distribution was only fitted to $\pi^*_{filt}$ and $\pi^*_{O3}$.

$$G_j = PDFfit(C^*_j) \text{ respectively } H_j = PDFfit(\pi^*_j) \qquad j = 1,\ldots,k \tag{4}$$

Where $G_j$ is the parameter pair of the PDF fitted to the concentration bootstrap sample $C^*_j$, $H_j$ is the parameter pair of the PDF fitted to the reduction bootstrap sample $\pi^*_j$, and *PDFfit* is the fit function in Matlab® for the chosen PDF type.

## Non parametric treatment model

Monte Carlo simulation was used to model reduction of pathogens by treatment. By using the bootstrap samples of $C^*$ and $\pi^*$ in Equation 5 a non-parametric model of *Campylobacter* reduction by treatment was achieved. This model was used to verify which of the methods (random, date or rank) provided the best validation for Monte Carlo simulations. The number of draws in one simulation $n$ was set to $m$ (the number of monitoring samples) to verify whether the model predicted the distribution of concentrations after treatment correctly. The validation was considered to be correct when the predicted concentrations after treatment overlapped the monitored concentrations.

$$C^*_{out\_ij} = C^*_{in[n.p1][k.p2]}\pi^*_{[n.p3][k.p4]} \qquad i = 1,\ldots,n \quad j = 1,\ldots,k \tag{5}$$

## Parametric treatment model

To predict the likelihood of rare events of high concentrations, Monte Carlo simulation with the parametric PDF's ($G$ for the raw water concentration and $H$ for the reduction) was applied as:

$$C^{\#}_{out\_ij} = PDFrnd(G_{[k.p1]})PDFrnd(H_{[k.p2]}) \qquad i = 1,\ldots,n \quad j = 1,\ldots,k \tag{6}$$

Where $C^{\#}_{out\_ij}$ is the predicted concentration after the treatment step, *PDFrnd* is the random draw of realisations from a given PDF function in Matlab®, $G_{[k.p1]}$ and $H_{[k.p2]}$ are a random PDF parameter pair of the raw water and the reduction respectively. The number of simulations $n$ was chosen with respect to the proportion of time that was of interest (i.e. $n = 10{,}000$ was applied in this study to predict events that can occur up to 0.01% of the time).

## Risk calculation

Exposure was calculated from the *Campylobacter* concentration in the drinking water and consumption of unboiled drinking water. For QMRA purposes the consumption can also be modelled as a PDF. However for this study only the average consumption was used since the goal was to show the impact of treatment modelling (using a PDF for consumption would distort these effects). Daily exposure (dose) $\mu_d$ (*Campylobacter*/d) was calculated by multiplying the estimated concentration with the average Dutch consumption of 0.177 litre of unboiled drinking water per day (Mons *et al.*, 2007). The daily risk of infection $P_{inf\_d}$ (infection per person per day) was calculated from exposure using a Beta-Poisson dose-response model for *Campylobacter* with $\alpha = 0.145$ and $\beta = 7.59$ (Medema *et al.*, 1996).

$$P_{inf\_d} \approx 1 - \left(1 + \frac{\mu_d}{\beta}\right)^{-\alpha} \qquad (\beta \geq 1 \text{ and } \alpha \leq \beta) \tag{7}$$

Since the concentration in the drinking water varies in time, the exposure also varies in time. In theory a frequently occurring low concentration could pose the same average yearly health risk as a rarely occurring high concentration. To assess the relative impact of occurring concentrations, the yearly risk of infection from exceeding a given concentration $C_i$ (organisms/L) for a proportion of the year $F_i$ (dimensionless) was calculated with Equation 8. Yearly risk of one or more infections $P_{inf\_y}$ (infection per person per year) was calculated with $F_i = 1$.

$$P_{inf\_y\_i} = 1 - (1 - P_{inf\_d\_i})^{365\,F_i} \tag{8}$$

# RESULTS

## Microbial monitoring

Samples were taken at the raw water sampling point, mixed filter effluent and ozonation effluent. *Campylobacter* was analysed monthly in 2003 and 2005. In the winter period (December to February) of 2003, 2004 and 2005

*Campylobacter* analysis was performed weekly. Table 5.1 provides an overview of the collected data. Figure 5.2 shows the sample results on a time scale including the uncertainty due to the MPN method.

**Table 5.1** Overview of *Campylobacter* monitoring data in Most Probable Number/L (MPN/L)

| | # | Mean MPN/L | Median MPN /L | Min MPN/L | Max MPN/L | St. Dev. | Skew | Kurtosis |
|---|---|---|---|---|---|---|---|---|
| Raw | 41 | 197 | 110 | 0.30 | 1,100 | 224 | 1.90 | 7.53 |
| Filtered | 32 | 11.6 | 4.0 | 0.40 | 110 | 20.6 | 3.72 | 17.65 |
| Ozonated | 31 | 0.04 | <0.03 | <0.03 | 0.40 | 0.10 | 3.26 | 12.26 |

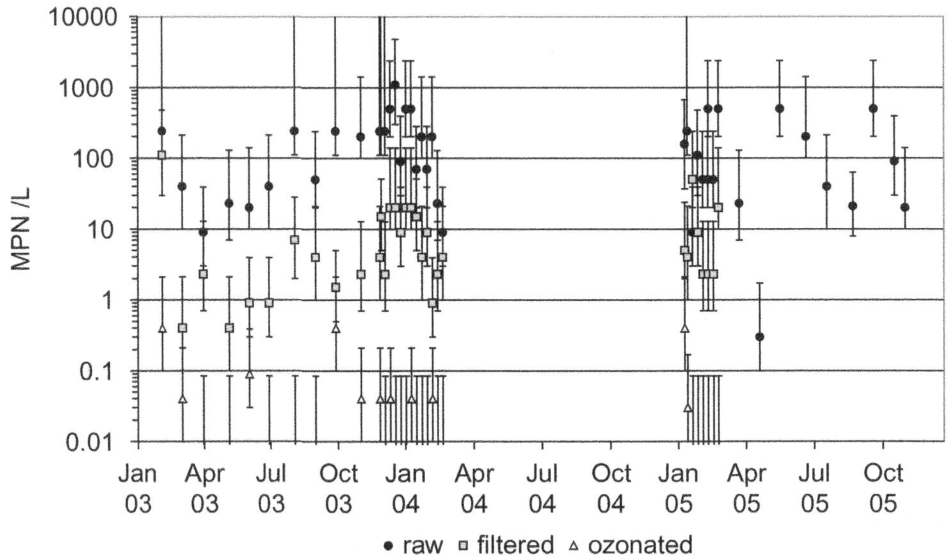


**Figure 5.2** *Campylobacter* monitoring results in raw water, filtered water and ozonated water. Error bars indicate the 95% confidence interval of the MPN for *Campylobacter*

## Methods to present distribution of concentrations

The variation of *Campylobacter* concentration in time needs to be taken into account for QMRA. Currently monitoring data is presented in QMRA studies as histograms to fit a PDF or cumulative histograms to fit a CDF on a semi-log scale. In this study the data was presented as Complementary Cumulative 'Histogram' to fit a Complementary Cumulative Distribution Function (CCDF) on a double log scale. This form of presenting data is generally applied in other fields of risk assessment and is well suited for extrapolation to rare events. Since the proportion of samples (similar to frequency) is plotted on log scale, and 'rare' events occur a small proportion of the time, this part of the data is 'magnified'. Figures 5.3a, 5.3b and 5.3c show the raw water monitoring data as PDF, CDF and CCDF respectively.

Figure 5.4 shows the CCDF of the monitored *Campylobacter* MPN numbers in raw, filtered and ozonated water. It also shows the median and 95% confidence interval, as determined with non-parametric MPN bootstrap. Since 30 to 40 *Campylobacter* samples were taken at each sampling point, each sample represents a proportion of 2.5–3.3% of the produced water.

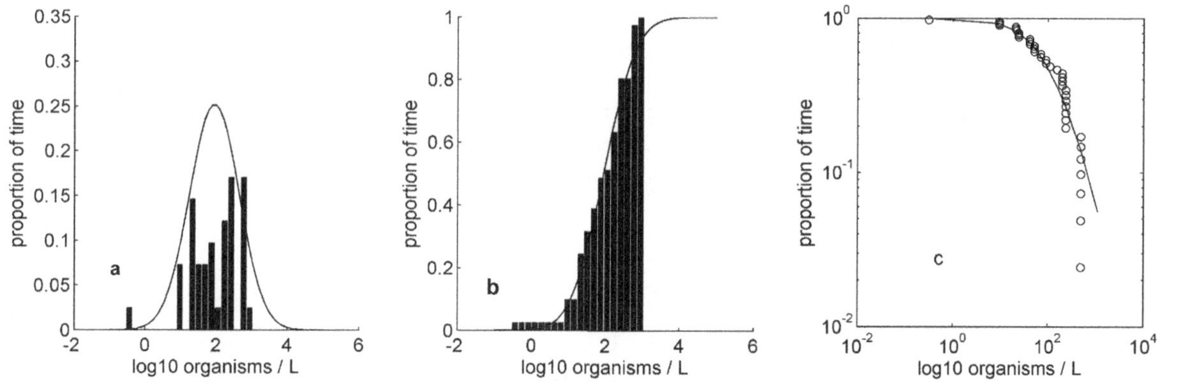


**Figure 5.3** Distribution of raw water *Campylobacter* monitoring data (bars or markers) and fitted lognormal distribution (lines) as PDF (5.3a), CDF (5.3b) and CCDF (5.3c)

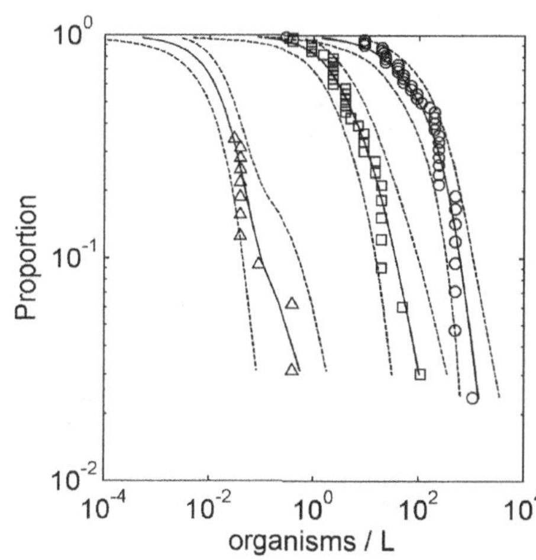


**Figure 5.4** CCDF of monitored *Campylobacter* MPN concentrations (markers) and the medians (lines) and 95% CI (dashed lines) of the non-parametric bootstraps for raw water (○), filtered water (□) and ozonated water (△)

## Non-parametric treatment model

The non-parametric stochastic model of treatment efficacy was validated with the *Campylobacter* monitoring data (Table 5.1 and Figure 5.4) by non-parametric validation of treatment efficacy. Figures 5.5a and 5.5b show the estimated *Campylobacter* removal by filtration $\pi^*_{filt}$ using the random, date and rank method. The removal found with the random method showed the highest variability of treatment efficacy. The date method resulted in similar removal, so pairing samples by date had little impact on the estimation of removal. Both the random and the date method results allowed for 'negative removal' to occur ($\pi_{filt} > 1$). This would imply that pathogens were sometimes 'produced' by the filter, which is unlikely. The rank method resulted in approximately 1 log removal and little variation. The rank method did not allow for negative removal.

Figures 5.6a and 5.6b show the estimated inactivation of *Campylobacter* by ozonation. The random and the date method resulted in a very similar estimate of inactivation by ozonation. Both showed high variability of inactivation and possible occurrence of negative inactivation. The rank method resulted in a more stable inactivation of approximately 2 log.

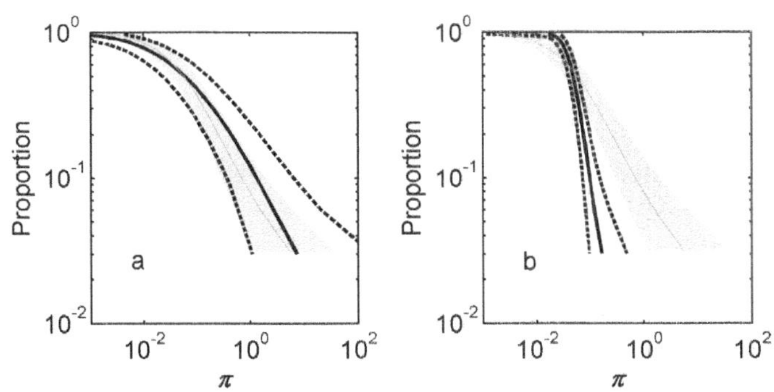


**Figure 5.5** Non-parametric validation of *Campylobacter* removal by filtration $\pi^*_{filt}$ with the date method (grey area) compared to random (5.5a) and rank (5.5b) method, median (line) and 95% CI (dashed)

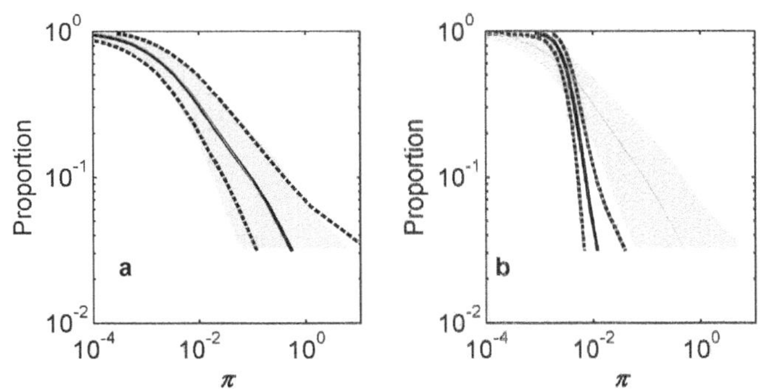


**Figure 5.6** Non-parametric validation of *Campylobacter* inactivation by ozonation $\pi^*_{O3}$ with the date method (grey area) compared to random (5.6a) and rank (5.6b) method, median (line) and 95% CI (dashed)

The estimated removal and inactivation in the previous section provided the non-parametric validation of the stochastic treatment model. The impact of the method of validation on the predicted concentrations after a treatment step was determined for the currently applied date method and the new rank method. The random method was not included in the rest of the study since the results were very similar to the date method. The concentrations after filtration were calculated from the raw water bootstrap samples (Figure 5.4) and the validated removal using the date method or the rank method (Figure 5.5b). The calculated concentrations were compared to the bootstrap of filtered water monitoring results $C^*_{filt}$ in Figure 5.7a (date method) and 5.7b (rank method). The concentrations after ozonation were calculated from the filtered water bootstrap samples (Figure 5.4) and the validated inactivation using the date method or the rank method (Figure 5.6b). The calculated concentrations were compared to the bootstrap of ozonated water monitoring results $C^*_{O3}$ in Figure 5.8a (date method) and 5.8b (rank method).

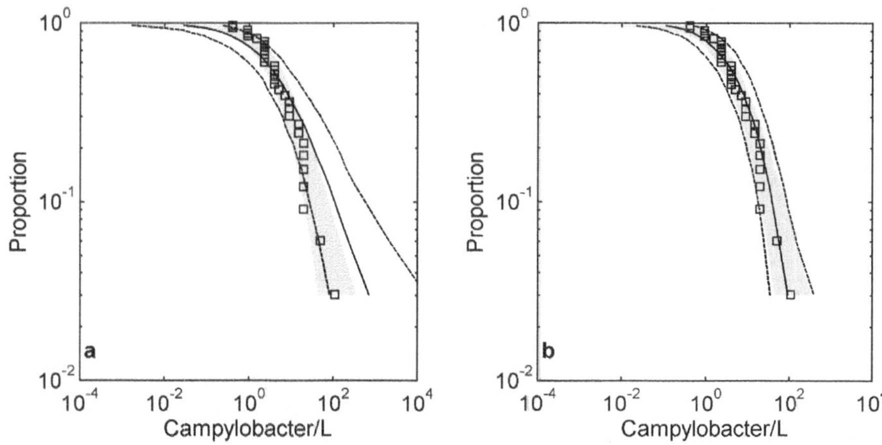


**Figure 5.7** *Campylobacter* concentration in filtered water calculated with the non-parametric model validated by conventional date method (5.7a) and new rank method (5.7b), median calculated concentration (line) and 95% CI (dashed), compared to monitored concentrations (markers) and 95% CI of $C^*_{filt}$ (grey area)

Figures 5.7a and 5.8a show that the date method resulted in substantial overestimation of *Campylobacter* concentrations both after filtration and ozonation. The rank method in provided an appropriate estimate of $\pi_{filt}$ and $\pi_{O3}$ for Monte Carlo simulation since the monitored concentrations in Figures 5.7b and 5.8b are in line with the predicted concentrations. The rank method was used in the rest of the study since it provided the best validation of the treatment model. The date method was included to demonstrate the error caused by this currently used method.

## Parametric treatment model

The non-parametric model cannot predict rare events of high *Campylobacter* concentrations or poor treatment removal due to the limited number of samples. The non-parametric model validations $C^*_{raw}$, $\pi^*_{filt}$ and $\pi^*_{O3}$ were therefore extrapolated with parametric distributions. Figures 5.9a and 5.9b shows that both the Weibull and gamma distributions underestimated rare high *Campylobacter* concentrations in raw water. This would result in underestimating the risk of infection from rare events. The lognormal distribution in Figure 5.9c matched the shape of $C^*_{raw}$ and was therefore chosen to extrapolate the raw water *Campylobacter* concentrations in this study.

The obtained bootstraps of reduction by treatment $\pi^*_{filt}$ and $\pi^*_{O3}$ were extrapolated to rare events of poor reduction (high values of $\pi$) with the Weibull, beta, gamma and lognormal distributions. Figure 5.10a shows the fit of the gamma distribution to $\pi^*_{filt}$. Weibull and beta distributions provided a practically identical graph and are therefore not shown.

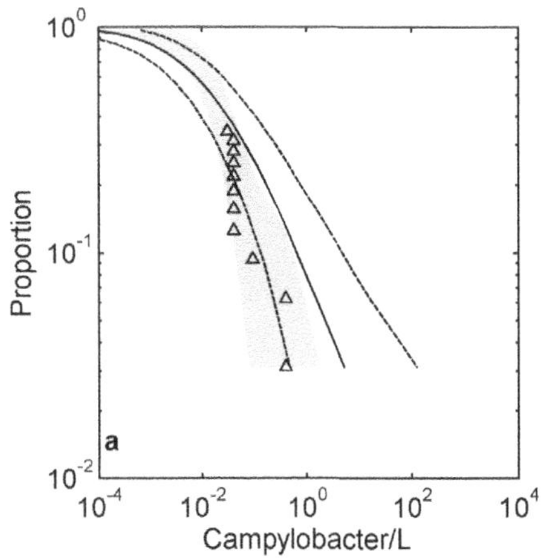


**Figure 5.8** *Campylobacter* concentration in ozonated water calculated with the non-parametric model validated by conventional date method (5.8a) and new rank method (5.8b), median calculated concentration (line) and 95% CI (dashed), compared to monitored concentrations (markers) and 95% CI of $C^*_{O3}$ (grey area)

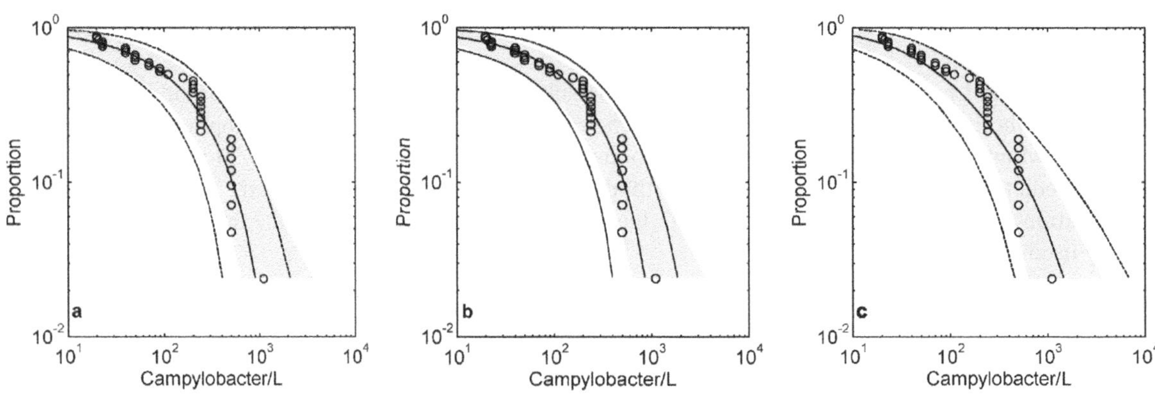


**Figure 5.9** Median (line) and 95% CI (dashed) of Weibull (5.9a), gamma (5.9b) and lognormal (5.9c) distributions fitted to the non parametric bootstrap (95% CI in grey) of *Campylobacter* concentrations in raw water. Markers indicate the monitored concentrations in MPN/L

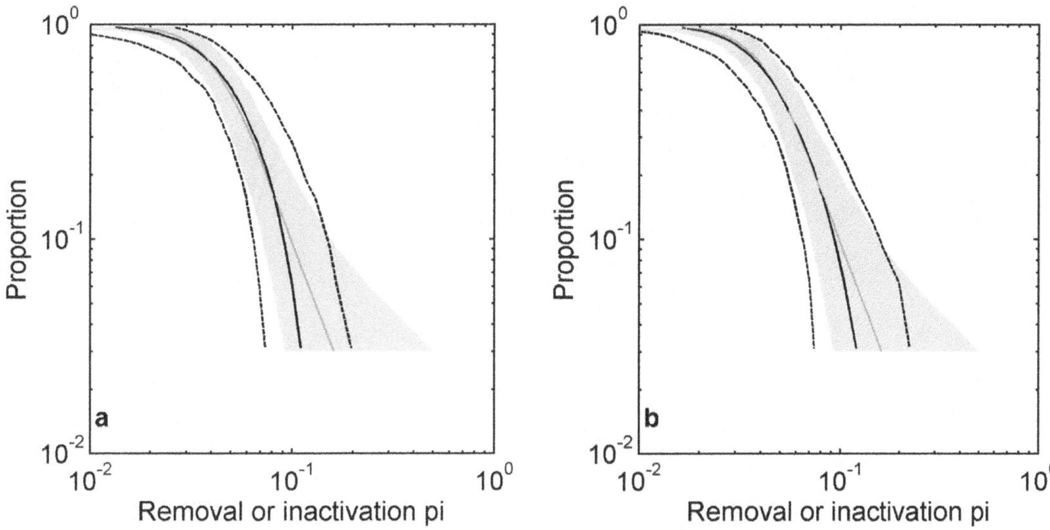


**Figure 5.10** Median (line) and 95% CI (dashed) of gamma (5.10a) and lognormal (5.10b) distributions fitted to the non parametric bootstrap of *Campylobacter* removal by filtration $\pi^*_{filt}$ (grey area indicates 95% CI, grey line indicates median). Weibull and beta distributions (not shown) provided a graph identical to the gamma distribution (a)

Figure 5.10b shows the fit of the lognormal distribution to $\pi^*_{filt}$. Although all distributions provided a reasonable fit for most of the data, only the lognormal distribution provided a reasonable fit to the high $\pi^*_{filt}$ values (poor removal). Therefore the lognormal distribution was used in further analysis in this study.

The gamma and lognormal distributions were fitted to the bootstrap of inactivation by ozone $\pi^*_{O3}$. Again the lognormal distribution provided the best fit to rare events of poor inactivation. The lognormal distribution was therefore used in the rest of the study.

## Parametric model of total chain

Monte Carlo simulation of the treatment from raw to ozonated water was performed to estimate the occurrence of *Campylobacter* in ozonated water. The parametric model was used to include normal rare events. The concentration in raw water, removal by filtration and inactivation by ozonation were modelled with lognormal distributions. Figures 5.11a and 5.11b shows the resulting median and 95% CI of predicted *Campylobacter* concentrations at each step compared to the monitored concentrations and their 95% CI. The date method shown in Figure 5.11a predicts concentrations after filtration and ozonation that are very high compared to the monitoring results. This indicates that currently applied QMRA methods based on pairing monitoring data by date can significantly overestimate the concentration of pathogens after treatment. The new method of pairing by rank resulted in a stochastic model of treatment that predicts concentrations that are in line with monitoring results (Figure 5.11b). Since the same data was used for validation and verification, this study only demonstrated that the rank method results in an accurate model, whereas the date method overestimated concentrations. The predictive accuracy of the rank method will be assessed in a subsequent study by using separate datasets for validation and verification.

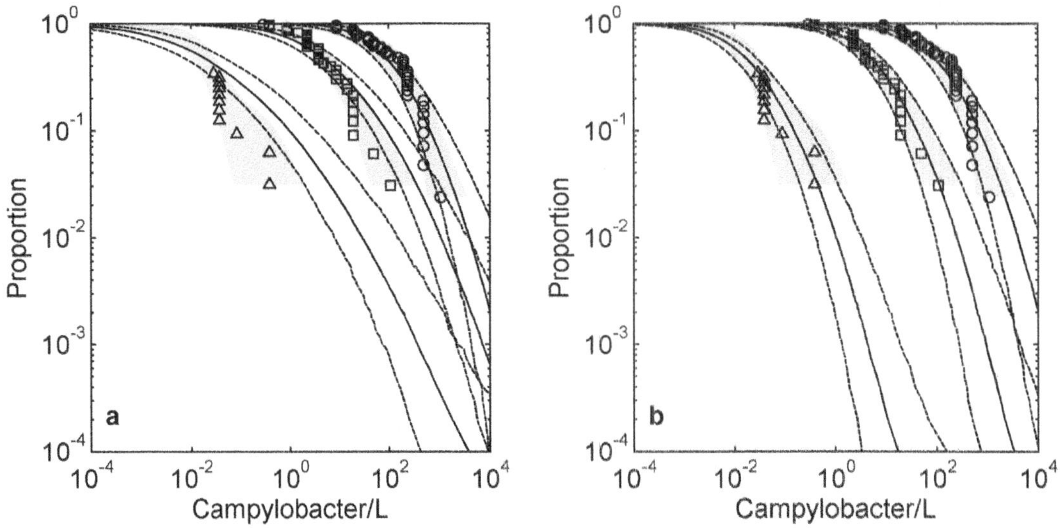


**Figure 5.11** Monte Carlo simulation of *Campylobacter* concentrations at different stages of treatment validated with the date (5.11a) and rank (5.11b) method, median (lines) and 95% CI (dashed), compared to monitored concentrations (markers) and 95% CI of monitoring (grey area) at several stages in treatment; raw water (○), filtered water (□) and ozonated water (△)

## Modelled risk of infection

The risk of infection from consuming ozonated water was calculated based on the modelled *Campylobacter* concentration in ozonated water. The choice of method to determine reduction by treatment had a significant impact on the assessed risk. The individual health risk is represented by the average yearly risk of infection. The date method predicted a 70% (33%–96%) average yearly risk of infection, whereas the rank method predicted 8.3% (3.8%–18%). So the conventional date method predicted a ten times higher average yearly risk of infection than the new rank method. The Dutch drinking water guidelines (Anonymous, 2001) require a maximum individual risk of $10^{-4}$ yearly average risk of infection, which corresponds to $2.75*10^{-7}$ daily risk of infection. Approximately 3 log reduction was needed in order to achieve this level of safety in the drinking water. The slow sand filtration at WTP Leiduin further treated the ozonated water to achieve this reduction.

Figure 5.12a shows that according to the date method the risk was dominated by concentrations of approx. 28 *Campylobacter*/L (black line) occurring in 1% of the water (grey line), which corresponds to an average yearly risk of 25% (black line). This concentration is 70 times higher than the maximum monitored concentration of 0.4 *Campylobacter*/L in Table 5.1 observed in 3% of the samples. Figure 5.11a however shows that the concentration after ozonation predicted with the date method is not in line with the monitoring data therefore this high estimate of risk seems unlikely.

Figure 5.12b shows that according to the rank method the average yearly risk was dominated by concentrations of approx. 0.14 *Campylobacter*/L (black line) occurring in 10 % of the water (grey line), which corresponds to an average yearly risk of 1.7% (black line). This concentration was exceeded in 10% of the monitoring samples; therefore the estimate of the frequency was regarded accurate. The extrapolation to rare events through modelling predicted that higher concentrations did not have a significant impact on the average yearly health risk.

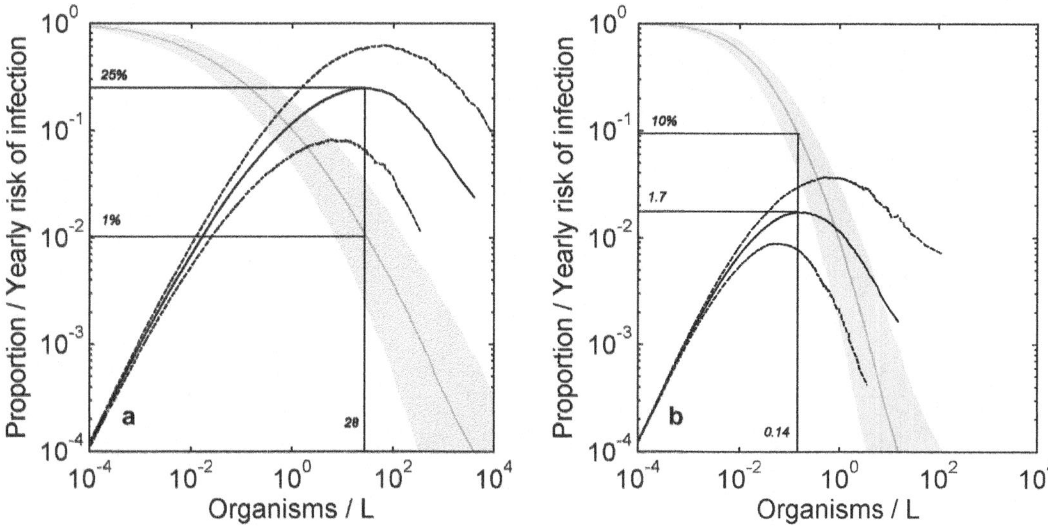


**Figure 5.12** Median (grey line) and 95% CI (grey area) of modelled *Campylobacter* concentration in ozonated water, and median (line) and 95% CI (dashed) of yearly risk of infection related to the proportion of each concentration using the date (5.12a) and the rank (5.12b) method

Since the choice of treatment model validation method appears to have a significant effect on the assessed risk, the model results need to be compared to the original monitoring data. The modelling also provides guidance for future monitoring. Frequent sampling of 10 L volumes will verify or improve the estimate of the concentrations that dominate the risk of infection. Lower concentrations which required larger volumes have little effect on the risk estimate. Smaller sample volumes would result in negative samples only, thus providing no additional information for the QMRA.

The FN curve of daily risk of infection is shown in Figure 5.13a and 5.13b. Filtration had a limited effect on the daily risk whereas ozonation had a major impact. The date method (Figure 5.13a) predicted more frequent occurrence of high risk than the rank method (5.13b). The rank method provided the best validation of the model, therefore only the FN curve for the rank method (Figure 5.13b) is discussed here. The FN curve shows both the variation of risk and the uncertainty of the assessed risk thus supporting decisions by risk managers and inspectors. The societal risk can be evaluated with the FN curve by evaluating the likelihood of simultaneous infection of a large number of people, referred to as an outbreak. An outbreak is represented in the FN curve by a high daily risk of infection. The FN curve in Figure 5.13b shows that the risk of infection from drinking ozonated water exceeds 0.7% one day per year (proportion of 0.0027). In a city of 1 million people 7,000 people would gain an infection of which some would develop illness. The upper 97.5 confidence limit of this estimate is 2% risk of infection one day per year, resulting in 20,000 infections. Since outbreaks may be detected when over 1% of the population becomes ill (Regli *et al.*, 1991), the risk assessment indicates that an outbreak might be detected yearly for this case study. A detected outbreak would result in a much greater effect on society than the incidental infections due to the yearly average risk. Current legislation does not set requirements for the acceptable frequency and magnitude of such an outbreak. The numbers in this example are hypothetical since the ozonated water passes slow sand filtration before distribution which reduces the risk.

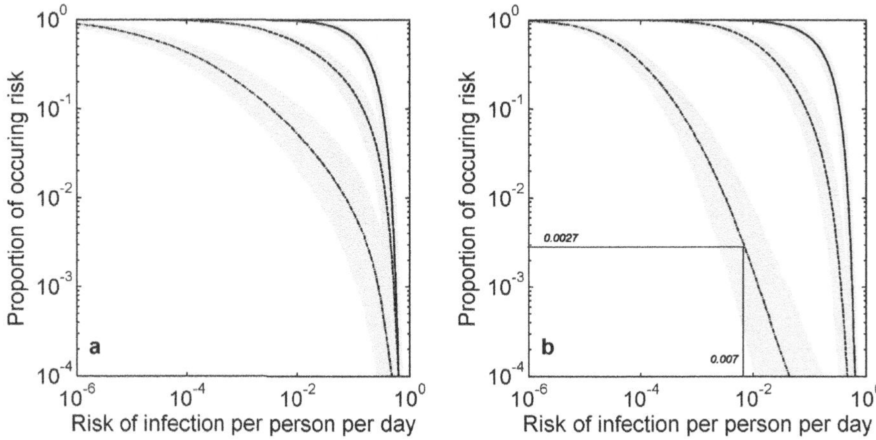


**Figure 5.13** FN curve of median (lines) and 95% CI (grey area) of daily risk of infection from drinking raw water (line), filtered water (dashed) or ozonated water (dash-dot) using the date method (5.13a) or the rank method (5.13b)

## DISCUSSION

Monte Carlo simulation of treatment is common practice in current QMRA studies. Since removal by treatment cannot be measured directly, it is calculated from concentrations before and after treatment measured on the same day. This approach assumes a correlation in time between these individual samples. However, it is known that such correlation is disturbed by several causes, even when the residence time in the treatment process is accounted for. Firstly, sampling variation due to (over-)dispersion of organisms in the water needs to be accounted for. Gale *et al.* (1997) showed that treatment enhances clustering of microorganisms, thus impacting the dispersion. Secondly, the residence time of particles in some processes (e.g. filtration) can be very different from the water residence time (Yao *et al.*, 1971). In addition, treatment processes vary in time (filtration cycles) and in space (inhomogeneous mixing of disinfectants). Finally microbial methods can have a large impact due to the quantification uncertainty (MPN, presence/absence) and recovery. Several methods have been published to account for these disturbances such as statistical correction for recovery (Teunis and Havelaar, 1999) or the use of copula's or correlations (Bukowski *et al.*, 1995 and Haas, 1999). In this case study, pairing by date resulted in the same assessed removal as random pairing, indicating that there is little correlation in time between influent and effluent data. The predicted concentrations based on the date method to estimate treatment removal efficacy did not match with the monitoring data *Campylobacter* MPNs. However the rank method applied in this chapter was in agreement with the monitoring data. The ramification from the rank method is that samples taken years apart may be paired, which contradicts to the intuitive expectation that only samples taken within a short time frame may be correlated. However, one must consider that the goal of the Monte Carlo simulation is to model the transition from the raw water distribution to the treated water distribution, not to predict the chance of an individual microorganism passing treatment.

The presented results were obtained for one case study, the applicability to other situations needs to be studied further. Since correlation in time may be relevant for other treatment systems, the choice of date or rank method must always be made with care. This study provided two methods for this. Firstly the random method provides a benchmark for data with no correlation. When the date method results in a significant deviance from the random method, this indicates that correlation in time has a significant effect. Secondly, concentrations after treatment predicted by non-parametric modelling should be in line with the validation data, taking into account the uncertainty of a limited dataset and method uncertainties.

Reported removal by treatment in literature is also applied in QMRA. Generally reported removal ranges over several log units, so the choice of removal in a QMRA study will significantly impact the assessed risk. One needs to consider that the date method was generally used to determine these literature values of removal. The results from this study lead to a new consideration of the reported data, since the rank method could lead to significant reduction of the range of reported removal.

The study adapted methods that are generally applied in other field of risk assessment, such as flooding, traffic or industrial accidents (Van Gelder, 1999), for application in drinking water for QMRA. This includes the use of CCDF, non-parametric bootstrapping and the FN curve. The main difference is that for many other fields of risk the extremes (water levels, fatal accident, earth quakes or process temperatures) can be monitored directly, leading to other extrapolation techniques such as peak over threshold (POT). It is unlikely that microbial monitoring catches the actual peak contaminations or moments of poor treatment. Microbial monitoring results must therefore be considered as random samples of the variation, to be extrapolated with statistical distributions. Adapting monitoring strategies to capture the real peaks may provide a significant improvement of the assessed risk. It must also be considered that the techniques presented in this study only predict the events due to (combinations of) 'normal' variations. Assessment of other 'man-made' events, such as operational errors or intentional contamination, need to be addressed with other methods, such as water safety plans (WHO, 2004).

Currently the individual risk, expressed as average yearly risk of infection or DALY, is the main parameter for risk evaluation (WHO, 2004). The prevention of outbreaks however is one of the main concerns of water utilities and health authorities. The FN curve allows for evaluation of both the individual risk and the societal risk of 'outbreaks'. Furthermore, it provides some insight into the uncertainty involved for both these aspects. The FN curve thus provides a basis for a new approach to risk evaluation and legislation.

Microbial monitoring remains important to verify the achieved level of safety. This study provided a method to determine the concentrations that are most relevant for the yearly average risk of infection. This can support monitoring programs in order to efficiently direct resources e.g. by taking frequent small volume samples, rather than a few large volume samples. Since the presented methods assume 'random' samples, a large volume sample cannot be considered as a large number of small samples. After all, it cannot be assumed that the distribution of concentrations in the large volume is identical to the distribution in the yearly produced water. Still large volume samples may be necessary to get a relevant number of positive samples per year. This means that sampling strategy may need to be adapted based on monitoring results: first find positives, and then determine concentrations most relevant for risk.

## CONCLUSIONS

The currently applied method to model drinking water treatment in QMRA was compared to an improved method. This study showed that the currently applied method did not predict the monitoring data used for validation in a case study with *Campylobacter* monitoring data of filtration and ozonation processes. Consequently the risk of infection was overestimated by one order of magnitude in this case study. The improved method accurately predicted the validation data. In this case the rank method proved to be the best validation method, however this may not be the case for all systems. The study also introduced other techniques to QMRA that improve calculation, presentation and evaluation of data and risk. Since CCDF graphs focus on rare events, visual evaluation of modelled extrapolation is improved. The use of non-parametric methods reduces the impact of PDF choice in an early stage of QMRA. Calculating the risk per concentration provides guidance for monitoring and the FN curve allows improved risk evaluation by distinguishing between individual and societal risk. Together these methods provide an improved protocol for modelling drinking water treatment in QMRA.

## REFERENCES


Anonymous (2001). Besluit van 9 januari 2001 tot wijziging van het waterleidingbesluit in verband met de richtlijn betreffende de kwaliteit van voor menselijke consumptie bestemd water. (Adaptation of Dutch drinking water legislation) *Staatsblad van het Koninkrijk der Nederlanden*, **31**, 1–53.

Bukowski J., Korn, L. and Wartenberg D. (1995). Correlated inputs in quantitative risk assessment: The effects of distributional shape. *Risk Analysis*, **15**(2), 215–219.

De Man J. C. (1975). The probability of most probable numbers. *European Journal of Applied Microbiology*, **1**, 67–78.

Gerba C. P. and Haas C. N. (1988). Assessment of risks associated with enteric viruses in contaminated drinking water. *ASTM Special Technical Publication*, **976**, 489–494.

Gale P., van Dijk P. A. H. and Stanfield G. (1997). Drinking water treatment increases microorganism clustering: The implications for microbial risk assessment. *Journal of Water Supply Research and Technology-Aqua*, **46**, 117–126.

Gibson III, C. J., Haas C. N. and Rose J. B. (1999). Risk assessment of waterborne protozoa: current status and future trends. *Parasitology*, **117**, 205–212.

Greenwood M. and Yule G. U. (1917). On the statistical interpretation of some bacteriological methods employed in water analysis. *Journal of Hygiene*, **16**, 36–54.

Haas C. N. (1983). Estimation of risk due to low doses of microorganisms: A comparison of alternative methodologies. *American Journal of Epidemiology*, **118**, 573–582.

Haas C. N. (1997). Importance of distributional form in characterizing inputs to Monte Carlo risk assessment. *Risk Analysis*, **17**(1), 107–113.

Haas C. N. and Trussel R. R. (1998). Frameworks for assessing reliability of multiple, independent barriers in potable water reuse. *Water Science & Technology*, **38**(6), 1–8.

Haas C. N., Rose J. B. and Gerba C. P. (1999). *Quantitative Microbial Risk Assessment.* Wiley, New York, USA.

Haas C. N. (1999). On modeling correlated random variables in risk assessment. *Risk Analysis*, **19**(6), 1205–1214.

Hijnen W. A. M, Veenendaal D., Van der Speld W. M. H., Visser A., Hoogenboezem W. and Van der Kooij D. (2000). Enumeration of faecal indicator bacteria in large water volumes using on site membrane filtration to assess water treatment efficiency. *Water Research*, **34**(5), 1659–1665.

Hrudey S. E. and Hrudey E. J. (2004). *Safe Drinking Water – Lessons Learned from Recent Outbreaks in Affluent Nations.* IWA Publishing, London.

ILSI Risk Science Institute Pathogen Risk Assessment Working Group (1996). A conceptual framework to assess the risks of human disease following exposure to pathogens. *Risk Analysis*, **16**(6), 841–847.

Medema G. J., Teunis P. F. M., Havelaar A. H. and Haas C. N. (1996). Assessment of the dose-response relationship of *Campylobacter Jejuni. International Journal of Food Microbiology*, **30**, 101–111.

Mons M. N., Van der Wielen J. M. L., Blokker E. J. M., Sinclair M. I., Hulshof K. F. A. M., Dangendorf F., Hunter P. R. and Medema G. J. (2007). Estimation of the consumption of cold tap water for microbiological risk assessment: An overview of studies and statistical analysis of data. *Journal of Water and Health*, **5**(S1), 151–170.

Nichols G. (2003). Using Existing Surveillance-Based Data. In: *Drinking Water and Infectious Disease: Establishing the Links*, Hunter P. R., Waite M., Ronchi E. IWA Publishing, London, UK.

Payment P., Berte A., Prévost M., Ménard B. and Barbeau B. (2000). Occurrence of pathogenic microorganisms in the Saint Lawrence River (Canada) and comparison of health risks for populations using it as their source of drinking water. *Canadian Journal of Microbiology*, **46**, 565–576.

Rose J. B., Haas C. N. and Regli S. (1991). Risk assessment and the control of waterborne *Giardia*sis. *American Journal of Public Health*, **81**, 709–713.

Regli S, Rose J. B., Haas C. N. and Gerba C. P. (1991). Modeling the risk from *Giardia* and viruses in drinking water. *Journal of the American Water Works Association JAWWA*, **83**(11), 76–84.

Smeets P. W. M. H., Van Dijk J. C., Stanfield G., Rietveld L. C. and Medema G. J. (2007). How can the UK statutory *Cryptosporidium* monitoring be used for quantitative risk assessment of *Cryptosporidium* in drinking water? *Journal of Water and Health*, **5**(S1), 107–118.

Teunis P. F. M., Havelaar A. H. and Medema G. J. (1994). *A literature Survey on the Assessment of Microbial Risk for Drinking Water*. RIVM Report 734301006, RIVM Bilthoven, the Netherlands.

Teunis P. F. M., Medema G. J., Kruidenier L. and Havelaar A. H. (1997). Assessment of the risk of infection by *Cryptosporidium* or *Giardia* in drinking water from a surface water source. *Water Research*, **31**(6), 1333–1346.

Teunis P. F. M. and Havelaar A. H. (1999). Cryptosporidium *in drinking water. Evaluation of the ILSI/RSI Quantitative Risk Assessment Framework*. Report No. 284 550 006 RIVM, Bilthoven, the Netherlands.

Teunis P. F. M., Evers E. G. and Slob W. (1999). Analysis of variable fractions resulting from microbial counts. *Quantitative Microbiology*, **1**, 63–88.

Teunis P., Davison A. and Deere D. (2004) *Short-term Fluctuations in Drinking Water Quality and Their Significance for Public Health*. Report, World health organization, Geneva, Switzerland.

Van Gelder P. H. A. J. M. (1999). *Statistical Methods for the Risk-Based Design of Civil Structures*. Thesis Delft University of Technology, the Netherlands.

Westrell T., Bergstedt O., Stenström T. A. and Ashbolt N. J. (2003). A theoretical approach to assess microbial risks due to failures in drinking water systems. *International Journal of Environmental Health Research*, **13**(2), 181–197.

WHO (2004). *Guidelines for Drinking Water Quality (Third edition)*. World health organization, Geneva, Switzerland.

Yao K. M., Habibian M. T. and O'Melia R. C. (1971). Water and waste water filtration: Concepts and applications. *Environmental Science & Technology*, **5**(11), 1105–1112.

# Chapter 6

# On the variability and uncertainty in quantitative microbial risk assessment of drinking water

*P.W.M.H. Smeets*[a,b], *P.H.A.J.M. van Gelder*[b], *J.C. van Dijk*[b] and *G.J. Medema*[a,b]

[a]*KWR Watercycle research institute, PO Box 1072, 3430 BB Nieuwegein, the Netherlands*
[b]*Faculty of Civil Engineering, Delft University of Technology, PO Box 5048, 2600 GA Delft, the Netherlands*

## ABSTRACT

Quantitative microbial risk assessment (QMRA) is increasingly applied to estimate drinking water safety. This study applied Monte Carlo (MC) modelling of drinking water treatment in QMRA using microbial monitoring data from a number of drinking water treatment plants to calibrate pathogen reduction. The study showed that pathogen concentrations after treatment could be predicted accurately when the model was properly calibrated. Current calibration methods pair influent and effluent samples by date to assess variability of reduction, however this resulted in overestimation of effluent concentrations by several orders of magnitude. An optimised calibration method was developed that, by definition, optimally predicted effluent concentrations. *Campylobacter* removal by filtration was predicted accurately using *E. coli* removal data to calibrate the MC model, even though direct comparison of data suggested no correlation between their removal. Application of the method to 20 datasets from various treatment sites provided a general overview of the efficacy and variability of full-scale treatment systems.

## INTRODUCTION

Since 1980 Quantitative Microbial Risk Assessment (QMRA) has been applied to quantify the microbial safety of drinking water (Haas, 1983; Gerba *et al.*, 1988; Regli *et al.*, 1991; Rose *et al.*, 1991; Haas *et al.*, 1993;

 *Stochastic Modelling of Drinking Water Treatment in Quantitative Microbial Risk Assessment*. By Patrick Smeets. ISBN: 9781843393740. Published by IWA Publishing, London, UK.

Teunis *et al.*, 1994; ILSI, 1996; Gibson III *et al.*, 1999 and Payment *et al.*, 2000). Risk of infection is calculated from the probability of ingesting pathogens (exposure or dose) and the probability of developing an infection from this exposure (dose-response relation) (Haas *et al.*, 1999). Pathogen concentrations in drinking water are generally below detection limits (Regli *et al.*, 1991). QMRA studies have therefore monitored pathogen concentrations in the raw water and modelled removal or inactivation by treatment to estimate concentrations in the drinking water (Teunis *et al.*, 1997; Haas and Trussel, 1998; Teunis and Havelaar, 1999; Medema *et al.*, 1999 and Westrell *et al.*, 2003).

Some studies used full-scale monitoring data of microorganisms before and after treatment steps to quantify treatment efficacy. A simple approach assumes a constant fraction of microorganisms in the influent reaches the effluent. The constant fraction is estimated from the monitoring data as the quotients of the arithmetic mean of effluent and influent concentrations (Teunis *et al.*, 1999). Variation of treatment efficacy has been observed in many studies. The variation has been accounted for in several studies by stochastically modelling treatment efficacy through Monte Carlo simulation. These studies generally paired influent and effluent samples taken on the same day to determine variation of treatment efficacy. A probability density function (PDF) was then fitted to this distribution e.g. through maximum likelihood estimation (Teunis *et al.*, 1997; Teunis *et al.*, 1999; Medema *et al.*, 1999; Masago *et al.*, 2004; Teunis *et al.*, 2004 and Smeets *et al.*, 2008). Estimation of the PDF parameters can be referred to as stochastic model calibration. The correctness of calibrated model could be validated by applying the model to the influent concentrations to predict the monitored effluent concentrations (Smeets *et al.*, 2008). The model was correctly calibrated when the monitored concentrations were within the credibility interval of the predict concentrations. Only few authors performed such validation. Medema *et al.* (1999) found that, when validating the model, the predicted concentrations after modelling a combination of treatment processes were higher than the observed concentrations used for calibration. They suggested that in reality interaction between treatment steps could reduce the variability leading to lower concentrations after treatment. Smeets *et al.* (2008) found that calibration by date did not result in an acceptably validated model even for a single process. They suggested that paring by date was not effective for calibration since over dispersion of microorganisms in water had a higher impact on observed concentrations that correlation in time between influent and effluent samples. Based on a single case study, Smeets *et al.* (2008) improved the methods for stochastic treatment model calibration. Validation of thus calibrated models resulted in an accurate prediction of microorganisms concentration after treatment. This study applied the improved methods developed by Smeets *et al.* (2008) to a range of other microorganisms and treatment systems in order to determine if these methods were generally applicable.

Several studies have used the monitored removal or inactivation of surrogate or indicator organisms at full-scale to predict reduction of pathogens (Teunis *et al.*, 1997; Teunis and Havelaar, 1999; Teunis *et al.*, 1999 and Medema *et al.*, 1999). Most used spores of sulphite-reducing Clostridia (SSRC) as a surrogate for *Cryptosporidium* reduction by treatment. However, many studies have found limited or no direct correlation between the reduction of pathogens and surrogate or indicator organisms (LeChevallier and Au, 2004, Hijnen *et al.*, 2004 and Hijnen *et al.*, 2005a, 2005b). None of the QMRA studies were able to verify if calibration with monitored indicator data provided an accurate model for pathogens reduction by treatment due to the lack of pathogen data. This study used both *E. coli* and *Campylobacter* treatment data to compare indicator and pathogen reduction.

Most QMRA studies need to rely on pilot tests reported in the literature to quantify the efficacy of treatment processes (LeChevallier and Au, 2004; Hijnen *et al.*, 2004 and Hijnen *et al.*, 2005a, 2005b). However, these pilot tests lack the variability and long-term effects of a full-scale situation. Variability can be caused by seasonal effects such as temperature, water quality effect such as increased turbidity after rainfall, operational effects such as backwashing one out of ten parallel filters, maintenance, operational changes such as dose adjustment and flow rate. Long-term effects at full-scale could be filter ripening, build up of particles in sedimentation and filtration over time, predation and inaccuracy of (dosing) equipment due to fouling or aging. In addition, many experiments to assess treatment efficacy apply a short-term spiked dose of microorganisms before treatment, which can affect the assessed efficacy due to differences in microbial populations (cultured or environmental) and lack of long-term effects. This study set out to compile data from different treatment situations at full-scale to provide a general overview of actual full-scale treatment performance.

Most QMRA studies have used information from the past (microbial monitoring) to predict future situations. Source water or treatment monitoring was typically performed for a period of one or a few years. However, conditions may change between years, even without being noticed. Thus the efficacy assessed in one year may not be applicable in another year. The year to year variability of a treatment process was therefore also studied.

The goal of the study was to verify that improved methods for stochastic treatment model calibration could provide more accurate validation than the pairing by date method. The improved methods were then used to evaluate the following applications of stochastic treatment modelling in QMRA of drinking water:

- Assessment of full-scale treatment efficacy and variability based on full-scale microbial monitoring;
- Stochastic treatment modelling to predict (variation of) microorganism concentrations after treatment and
- Calibration of a stochastic treatment model for *Campylobacter* removal using *E. coli* full-scale monitoring.

# METHODS

## Statistical methods

The stochastic treatment analysis and model used of Monte Carlo simulation of influent water microorganism concentration $C_{in}$ times reduction by treatment $\pi$. Both $C_{in}$ and $\pi$ were described by a probability density function (PDF) (Teunis *et al.*, 1997 and Smeets *et al.*, 2008). Reduction by treatment $\pi$ was assessed from monitored concentrations of microorganisms before and after full-scale treatment processes. Several methods for the calibration of $\pi$ were compared in this study. The mean in/out method (or 'constant fraction' method) used the quotients of the arithmetic mean of effluent and influent concentrations as a point estimate of treatment efficacy $\pi$ (Teunis *et al.*, 1999). The date method (pairing influent and effluent samples by date) was the current method to assess variable treatment efficacy (Teunis *et al.*, 1999 and Smeets *et al.*, 2008). The rank method (pairing influent and effluent samples by ranked concentration) had improved validation in a case study and the random method assumed no correlation between influent and effluent samples (Smeets *et al.*, 2008).

The optimised method to calibrate $\pi$ was developed in this study. A simple calibration routine was used to determine the optimal parameters of the PDF of $\pi$ (reduction by treatment). Parameters were considered optimal when validation of the model with the calibration data was accurate e.g. the monitored concentrations in the effluent were within the confidence interval of the predicted concentrations. This calibration routine used the parameters of the PDF of $\pi$, derived with the rank and the date method respectively, as initial estimates of the optimal PDF. The model was calibrated using parameter combinations incrementing in 50 steps from the rank to the date parameters pairs (resulting in 2,500 parameter combinations). The model was validated for each of these parameter pairs using 1,000 realisations of $m$ samples in the Monte Carlo simulation. Predicted concentrations were compared with the $m$ monitored concentrations after treatment. Thus 2,500 runs of 1,000 realisations of $m$ samples were performed. The optimal parameter pair was determined by a least squares method comparing the median of the PDF of predicted concentrations after treatment to the median of the PDF of monitored concentrations. The best fit was confirmed by visual evaluation of the predicted and monitored PDF. If the visual evaluation did not provide a satisfactory fit, the initial estimates were adapted to include a broader range of parameter pairs until a satisfactory fit was achieved.

Lognormal, gamma, Weibull and beta distributions were tested as PDF type to describe variation of treatment efficacy $\pi$. Gamma, Weibull and beta distributions generally did not provide a satisfactory fit of the tail of the distributions of poor treatment efficacy (high $\pi$ values). Therefore only the lognormal distribution was applied in this study.

Data analysis methods were described in (Smeets *et al.*, 2008) and included non-parametric hierarchical bootstrapping for data analysis to include microbial method uncertainty caused by most probable number (MPN) data. However, most microbial data in this study was quantified by direct colony count. The bootstrapping method was adapted for application to direct count data, in which case the likelihood $q$ of concentration $C$ given count $N$ was calculated according to a Poisson probability density function. The uncertainty in the concentration can be conveniently assessed by Bootstrap simulations.

In one example in this study, monitoring data was simulated to demonstrate what variability of $\pi$ would be observed based on yearly monitoring even if $\pi$ did not change between years. A single MC simulation of concentrations after treatment was performed using $m$ monitored concentrations in raw water of that year and $m$ random draws from the treatment PDF of $\pi$. This provided a simulated PDF of microorganisms in treated water for that year. Monitoring results for treated water in that year were simulated by randomly drawing $m$ samples from the simulated PDF.

The variable reduction by treatment was reduced to a single log-reduction in Table 6.1 for ease of presentation. The reduction was calculated as the $-\log_{10}$ of mean $\pi$ since this better represented total log reduction over a period of time than the mean of the log reductions.

## Monitoring data

Microbial monitoring data from catchment to tap systems (CTS) was collected by water companies and organisations within the MicroRisk project and by other water companies participating in QMRA research. Each water company used its own (standard) methods for microbial analysis. Data from Teunis and Havelaar (1999) was also evaluated. Table 6.1 provides an overview of the number of data per system, treatment process and organism.

# RESULTS

## Treatment performance assessment

Data from all the treatment systems was analysed with the conventional date method (pairing samples by date), the simple mean in/out method and with the random, rank and optimised method. Goal was to compare the estimates of $\pi$ of these different calibration methods. The reduction of each monitored microorganism was assessed for each treatment step of each CTS in Table 6.1. Since it was impractical to present all details of the assessments in this paper, some selected examples were used to illustrate the general findings. Figure 6.1 provides an example of monitored *E. coli* concentrations before and after filtration. *E. coli* were clearly reduced by treatment. The steepness of the distribution curve was slightly reduced by filtration, indicating that variability of *E. coli* concentrations slightly increased due to filtration.

**Table 6.1** Results of the treatment assessments based on microbial monitoring data collected in the MicroRisk project using the current mean in/out or pairing by date method versus the improved optimised method. SSRC=spores of sulphite reducing *Clostridia*; *E.coc* = *Enterococcus*; *E. coli* = *Escherichia coli*; Tot. cf = Total coliforms; *Camp.* = *Campylobacter*. GAC = granular activated carbon

| | | # Sample influent | Mean influent conc. #/L | # Samples effluent | Mean effluent conc. #/L | Log $\frac{mean_{out}}{mean_{in}}$ | Date Method | Improved method (optimised lognormal PDF) | | | | |
|---|---|---|---|---|---|---|---|---|---|---|---|---|
| | | | | | | | Log mean π | Log mean π | Log median π | Log 97.5% π | π PDF μ | π PDF σ |
| **CTS1** | | | | | | | | | | | | |
| Pre-oxidation | SSRC | 31 | 381 | 29 | 289 | 0.12 | 0.07 | 0.13 | 0.23 | −0.35 | −0.52 | 0.68 |
| | *E. coc* | 27 | 5,349 | 22 | 509 | 1.02 | 0.81 | 1.02 | 1.14 | 0.50 | −2.63 | 0.75 |
| Clarification | *E. coli* | 91 | 603 | 58 | 70 | 0.93 | 0.41 | 0.51 | 1.11 | −0.30 | −2.56 | 1.66 |
| | SSRC | 29 | 289 | 29 | 7.35 | 1.59 | 1.20 | 1.64 | 1.70 | 1.27 | −3.91 | 0.50 |
| Filtration | *E. coli* | 58 | 70 | 58 | 9.31 | 0.88 | 0.00 | 0.84 | 0.91 | 0.43 | −2.09 | 0.56 |
| | SSRC | 29 | 7.35 | 29 | 4.35 | 0.23 | 0.73 | 0.40 | 1.31 | −0.44 | −3.02 | 2.05 |
| Inter-Ozonation | SSRC | 29 | 4.70 | 29 | 0.71 | 0.82 | −0.80 | 0.35 | 0.48 | −0.18 | −1.11 | 0.78 |
| GAC filtration | SSRC | 29 | 0.71 | 16 | 0.16 | 0.65 | −0.30 | 0.44 | 0.44 | 0.41 | −1.02 | 0.04 |
| **CTS2** | | | | | | | | | | | | |
| Filtration | *E. coli* | 1061 | 260 | 863 | 12.5 | 1.32 | 1.10 | 1.28 | 1.50 | 0.64 | −3.46 | 1.01 |
| | SSRC | 507 | 494 | 451 | 16.3 | 1.48 | 1.24 | 1.54 | 1.67 | 1.02 | −3.85 | 0.77 |
| Ozonation | *E. coli* | 863 | 12.5 | 1166 | 0.02 | 2.76 | 1.94 | 2.68 | 2.82 | 2.13 | −6.49 | 0.81 |
| | SSRC | 451 | 16.3 | 470 | 2.52 | 0.81 | 0.39 | 0.85 | 0.97 | 0.33 | −2.24 | 0.75 |
| Slow sand filtration | SSRC | 470 | 2.52 | 1394 | 0.15 | 1.23 | 0.36 | 1.16 | 1.17 | 1.04 | −2.69 | 0.15 |
| **CTS10** | | | | | | | | | | | | |
| Pre-oxidation | *E. coli* | 13 | 621 | 8 | 15.3 | 1.61 | 0.48 | 1.51 | 1.54 | 1.24 | −3.54 | 0.35 |
| | *E. coc* | 13 | 179 | 8 | 2.09 | 1.93 | 0.42 | 1.48 | 1.53 | 1.11 | −3.53 | 0.50 |
| | Tot cf. | 13 | 2,876 | 8 | 163 | 1.25 | 0.44 | 0.67 | 1.47 | −0.16 | −3.38 | 1.91 |
| **CTS12** | | | | | | | | | | | | |
| Filtration | *E. coli* | 82 | 77.6 | 288 | 9.90 | 0.89 | 0.61 | 0.94 | 1.28 | 0.21 | −2.95 | 1.26 |
| Ozonation | *E. coli* | 288 | 9.90 | 296 | 0.73 | 1.13 | 0.14 | 0.81 | 1.24 | 0.04 | −2.86 | 1.41 |
| **CTS13** | | | | | | | | | | | | |
| Slow sand filtration | *Camp.* | 289 | 10.01 | 735 | 0.01 | 3.07 | 1.34 | 3.13 | 3.35 | 2.50 | −7.71 | 1.00 |
| **Teunis and Havelaar** | **1999** | | | | | | | | | | | |
| Sedimentation-filtration | SSRC | 73 | 209.26 | 73 | 0.53 | 2.60 | 1.90 | 2.37 | 2.88 | 1.58 | −6.64 | 1.53 |

Several methods were applied to assess treatment efficacy based on the data in Figure 6.1. The currently applied date method assumed a correlation in time between influent and effluent samples. However, when influent and effluent data were paired randomly (random method) the assessed reduction $\pi$ was almost identical to reduction assessed through pairing by date, as was illustrated in Figure 6.2a. This illustrates that pairing by date did not affect the calibration of $\pi$. The rank method assumed that influent and effluent concentrations were correlated by concentration, e.g. that a high influent concentration would at some moment in time lead to a high effluent concentration. Pairing samples by rank led to relatively little variation of reduction, as illustrated in Figure 6.2b.

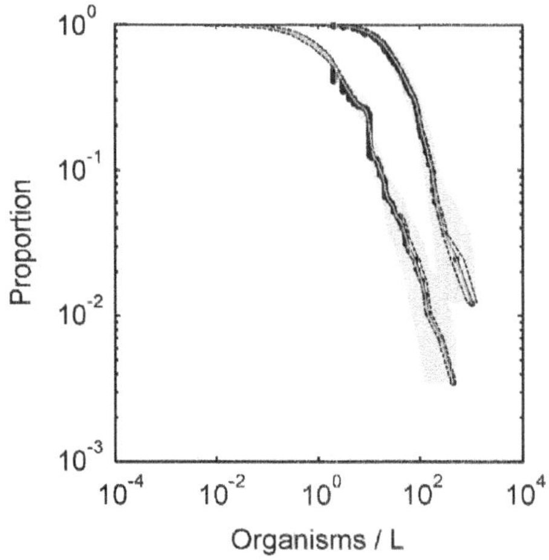


**Figure 6.1** Monitored *E. coli* concentrations before and after filtration at CTS 12 (markers), 95% CI of the individual concentrations (dashed lines) and 95% CI of the total dataset assessed through bootstrapping (grey area)

The assessed distributions of reduction by the date and the rank method in Figure 6.2a and 6.2b were used in the stochastic model to predict the concentrations after filtration based on influent concentrations. Lognormal probability density functions (PDF) were fitted to the influent concentration $C_{in}$ and the assessed reduction $\pi_{date}$ and $\pi_{rank}$ respectively. These were then applied in the stochastic model by Monte Carlo simulation to predict $C_{out}$.

Figure 6.3a shows that the date method significantly overestimated the concentrations after treatment. The rank method in Figure 6.3b slightly underestimated concentrations after treatment. Since the same data were used in the assessment to determine $\pi$, the predicted concentrations should perfectly match the monitored data. This suggested that both assumptions of correlation by date or by rank were incorrect. Teunis *et al.* (1999) already predicted that the real distribution of $\pi$ would lie between these two extreme assumptions. The optimised approach was used to determine the PDF parameters of reduction by filtration $\pi$. Figure 6.3c shows that the monitored concentrations lie within the confidence interval of the predicted concentrations. Thus the correct calibration of the stochastic model was validated for the calibration data. Validation of the calibrated model for other datasets will be discussed later in this paper.

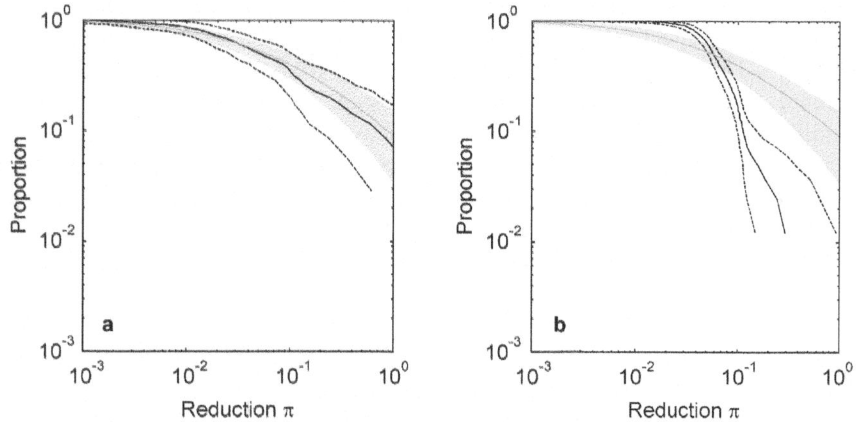


**Figure 6.2** Assessed reduction of *E. coli* by filtration at CTS 12 according to the random method (Figure 6.2a and 2b grey area), the date method (Figure 6.2a lines) and the rank method (Figure 6.2b lines). Dashed lines indicate the 95% credibility interval assessed through bootstrapping

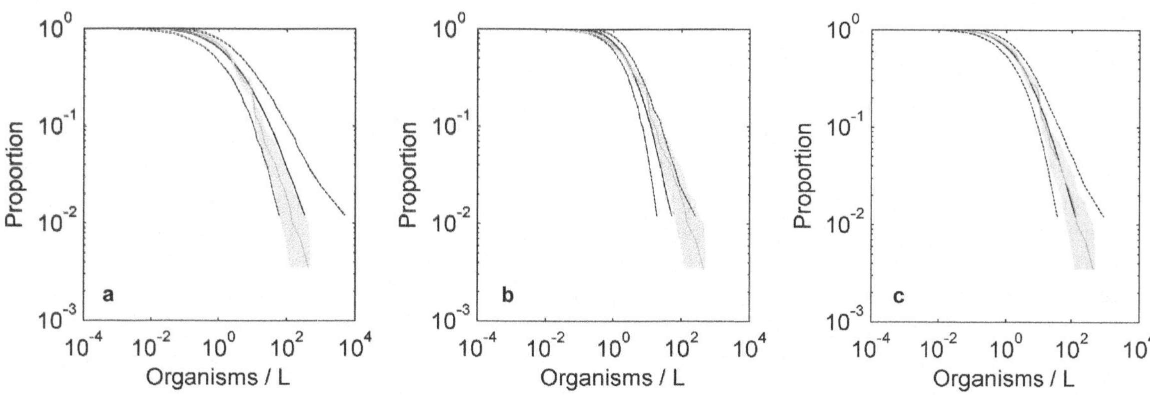


**Figure 6.3** Monitored *E. coli* concentrations after filtration at CTS 12 (grey area) compared to predicted concentrations using the date (6.3a), rank (6.3b) and optimised method (6.3c)

Table 1 provides an overview of the available data from the assessed treatment systems, and the results of the treatment assessment according to the date and the optimised method. The lognormal distribution provided the best fit of both monitored concentrations and assessed reduction $\pi$ in all the assessments with both methods. The variable reduction by treatment was reduced to a single log-reduction in Table 1 for ease of presentation. Table 1 will be discussed for the optimised, date and mean in/out methods separately.

## Optimised method

The optimised method was most effective to describe the transition of microorganism concentrations from influent to effluent of a treatment step. Therefore the assessed $\pi$ provides information on the treatment process efficacy and variability. Some processes were assessed for the reduction of the same organism at various locations. *E. coli* reduction by filtration sites CTS1, CTS2 and CTS12 was quite similar with averages of 0.84, 1,28 and 0.94 log reduction respectively. Filtration at CTS12 was more variable, with a 97.5% percentile of only 0.21 log, whereas CTS1 and CTS2 still provided 0.43 and 0.64 log reduction. Disinfection processes showed more differences between sites, which was expected since sites might apply different doses and operate under different conditions. Ozonation achieved 2.68 log inactivation of *E. coli* at CTS 2 whereas CTS 12 only achieved 0.81 log inactivation. Comparing the mean, median and 97.5% percentile of the optimized estimate of $\pi$ in Table 1 shows that some processes were fairly stable (e.g. *E. coli* inactivation by ozonation at CTS2) whereas other were more variable (e.g. *E. coli* inactivation by ozonation at CTS12). Comparing the optimized estimate of mean $\pi$ in Table 6.1 to reductions reported in literature (Hijnen, 2005a, 2005b) shows that treatment processes at full-scale generally appeared to achieve the lower range of reported reduction.

The optimised method proved to be the most effective method to determine parameters of the reduction PDF for the purpose of stochastic modelling. The rank method and the optimised method often provided very similar results that accurately predicted effluent concentrations. Visual evaluation of the results was very effective to detect errors in data and provided a very intuitive measure for goodness of fit compared to statistical parameters. The results of the optimised method were compared to results of other methods that are currently used; the mean in/out method and the pairing by date method. The accuracy of these methods could thus be evaluated. Secondly the optimised method was applied to test how well stochastic modelling was able to predict reduction of microorganisms.

## Currently applied date method

All systems in Table 6.1 were also assessed with the date method. Evaluation of the assessed reduction (similar to Figure 6.2) led to the conclusion that the date and random methods generally provided very similar estimates of reduction $\pi$. High variability of treatment efficacy was thus observed. Apparently the overdispersion of microorganisms in water dominated the observed concentrations such that pairing observations by date was no longer valid. The high variability led to underestimation of treatment efficacy. The level of underestimation depended on the site, such as the number of data points and the shapes of their distributions. The assessed reduction with the date method was compared to the optimised method in Figure 6.4. In all but one cases the date method underestimated treatment efficacy, sometimes by two to three orders of magnitude.

Underestimation of treatment efficacy and overestimation of variability led to overestimation of effluent concentrations (similar to Figure 6.3a). Especially rare events of high effluent concentrations were much higher than would be expected based on the monitoring results. Figure 6.5 shows that in some cases this effect was very significant.

## Mean in/out method

The simplified approach of calculating $\pi$ from mean influent and effluent concentrations was compared to the statistical approach. Results are shown in Table 6.1 and Figure 6.6. The simplified approach provided a fairly accurate estimate of the mean reduction in 60% of the cases. However, it significantly overestimated treatment efficacy in 30% of the cases compared to the optimised method. In one case the simplified approach underestimated treatment efficacy. Remarkably the mean in/out method was more accurate than the date method. The mean in/out method could be used as a first estimate of treatment efficacy. However, it provides no information on treatment variability, and there is a risk of overestimating treatment efficacy. The optimised stochastic method should be preferred over the mean in/out method for QMRA since it provides more information based on the same data. The optimised method provides a more accurate estimate of treatment efficacy and the information on treatment variability is of importance to determine the likelihood and impact of events. Information on variability can also be of importance for treatment optimisation.

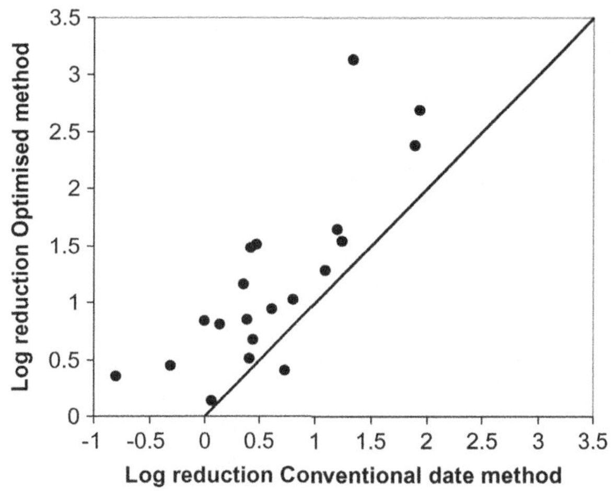


**Figure 6.4** Mean log reduction assessed with the conventional date method (pairing samples by date) versus assessment through the optimization method presented in this study

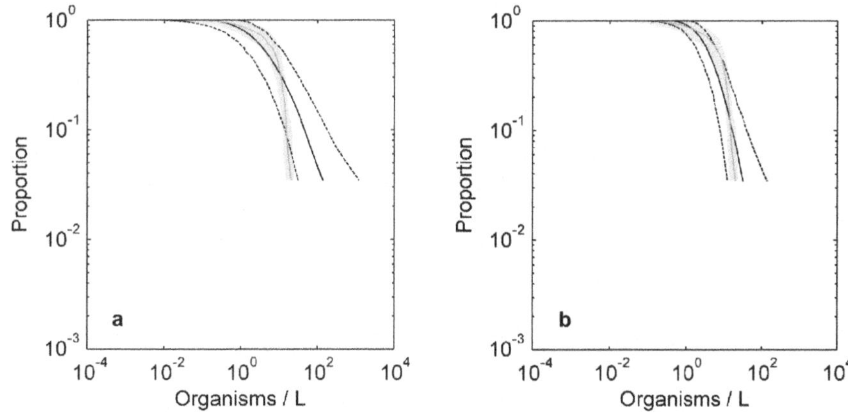


**Figure 6.5** Monitored (grey area) and predicted (lines) SSRC concentrations after clarification at CTS1 predicted with the date (6.5a) and the optimized (6.5b) method

Figures 6.5 and 6.6 compared the methods to calibrate $\pi$ from monitoring data. Calibration with the mean in/out and date methods generally did not result in a validated model. The model calibrated with the optimised method was consistently validated. This validation was performed with the monitoring data used for calibration in order to compare the methods. However, actual validation of the model requires application of the calibrated model to a different dataset. The goal of the development of a stochastic model was to predict concentrations that were not monitored and to quantify the uncertainties around the predictions. Therefore the optimised method was used in the rest of the study to test the validity and accuracy of stochastic treatment modelling in several applications.

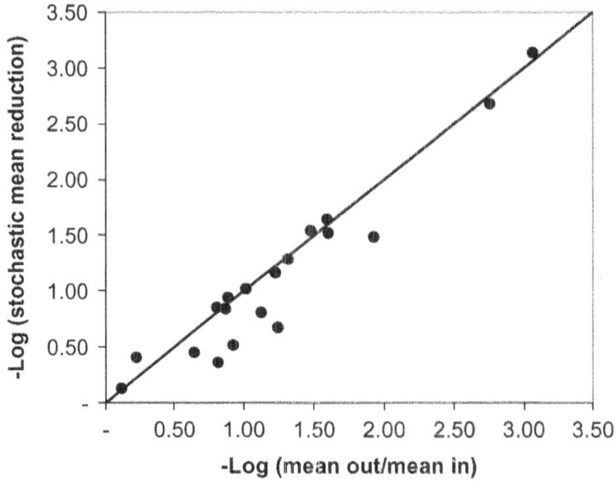


**Figure 6.6** Mean treatment efficacy of all cases determined by mean in/out method versus the optimised method

## Yearly variability of treatment performance

First the optimised method was used to assess treatment variability. The use of historical data to predict reduction of microorganisms by treatment assumes that reduction as it occurred in the past remains similar in the future. A dataset from CTS2 of ten years weekly samples was analyzed to determine how much variation in reduction occurs between subsequent years. Figure 6.7a shows the reduction of SSRC by filtration over a 10 year period. This shows that the level and variability of removal appeared to vary between years. The observed mean reduction varied between 1.2 and 1.7 log depending on the year that was monitored. During the year, reduction variation ranged over 0.6 to 2.5 log, depending on the assessed year. The analysis does not show whether the treatment efficacy actually varied between years, or the treatment did not vary and the observed variations were a consequence of monitoring effects.

A simulation was performed to determine whether the observed variation in Figure 6.7a was due to actual variations in treatment efficacy or due to sampling error. The PDF of reduction $\pi$ for the total 10 year period was used to generate 10 years of weekly monitoring data (52 samples per year). These simulated data were then analysed identically to the monitored dataset. Figure 6.7b shows that in this simulation, where the reduction was identical each year, the observed mean reduction varied between 1.3 and 1.8 log.

The range of variation within a year of 0.3 to 1.5 logs for the simulated data appears to be less than for the original data in Figure 6.7a. The original data showed more variation towards high removal than the simulated data. However these high removals are relatively unimportant in the risk assessment, since the lower removals dominate the risk. This simulation demonstrated that yearly observed variations of mean treatment efficacy could be caused by sampling effects. The difference in observed variation within a single year suggests that removal by filtration was not completely constant over the years. Operation of the filtration was not intentionally changed. However, unintentional variations in process conditions such as water temperature, pH and suspended solids concentration might have occurred which impacted treatment efficacy. A more detailed study might reveal correlations between these parameters and treatment efficacy, however, this information was not available for this study and is recommended for a follow-up study.

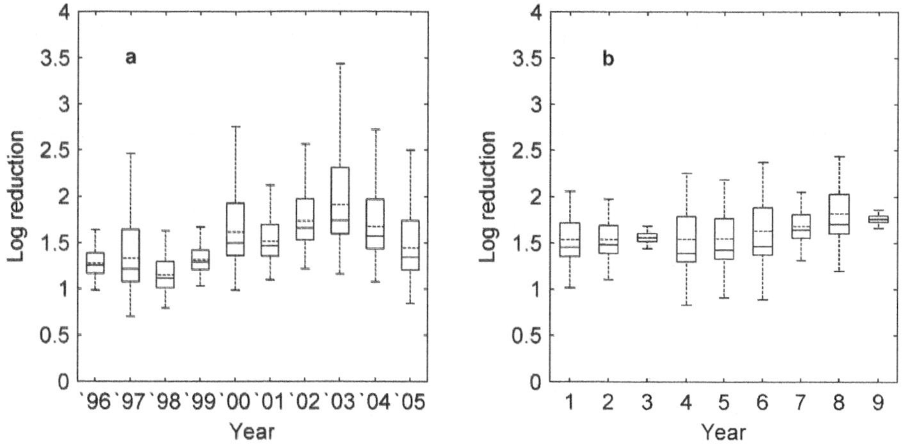


**Figure 6.7** Assessed reduction of SSRC by filtration at CTS 2 over a ten year period based on weekly monitoring of raw and filtered water (6.7a) and on simulated monitoring data using $\pi$ assessed for the ten year period (6.7b). Whiskers: 95% CI, box: 50% CI, line: mean reduction, dashed: median reduction

## Validity of the calibrated stochastic treatment model

The CTS 2 dataset of 10 years monitoring of SSRC before and after filtration was split up to test the validity of the calibrated stochastic treatment model. The data consisted of 52 samples before and after treatment per year. The first 5 years of data were used to validate the treatment model. This treatment model was then applied to the raw water SSRC data of each individual year, including the second 5 year period that was not used for validation. The predicted concentrations after filtration were then compared to the monitoring results. Figure 6.8 shows the monitored and predicted SSRC concentrations after filtration for each year. The predicted mean concentrations are very much in line with the monitored concentrations, both in the period used for validation and the following period. The model also predicted the yearly range of variation of these concentrations. The predicted concentrations change per year along with the monitored concentrations, indicating that the yearly change in source water concentration was the main cause for changes in observed post-filtration concentrations. Some inaccuracy was observed which may be due to sampling error. This example showed that stochastic treatment modelling can provide a realistic estimate of post treatment microorganism concentrations and their variability.

## Use of surrogate organisms

Water companies generally monitor for indicator organisms. When the indicator and pathogen reduction by treatment is expected to be similar, the observed reduction of the indicator organism could be applied to the pathogen. This was tested using data on *E. coli* and *Campylobacter* before and after filtration at CTS 2. Figure 6.9 shows the relation between directly observed *E. coli* and *Campylobacter* reduction when compared by date. Figure 6.9 suggested that there was little correlation between their reduction. Microbial method uncertainty for *Campylobacter* was significant since it was quantified in a 3X3 MPN (Most Probable Number) matrix (whiskers in Figure 6.9). Still the method uncertainty could not fully account for the lack of correlation between *E. coli* and *Campylobacter*. Applications of the date method as discussed before showed that pairing influent and effluent samples by date was inappropriate to quantify treatment efficacy. Correlation between reduction of two organisms therefore could not be determined by pairing by date either, so directly observed correlation should not be expected. Still the level and variation of *E. coli* reduction could be an indicator of *Campylobacter* reduction when the datasets were compared as a whole in a stochastic assessment.

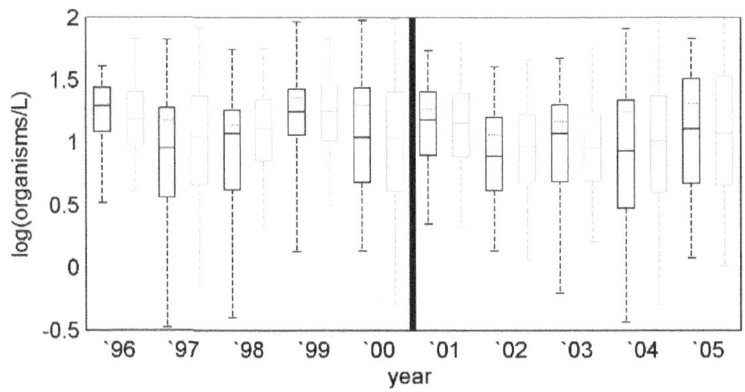


**Figure 6.8** Box-whisker plot of monitored (black) and predicted (grey) SSRC concentrations after filtration at CTS 2 per year. The model was calibrated with the first 5 years of data. Whiskers: 95% CI, box: 50% CI, line: median concentration, dashed: mean concentration

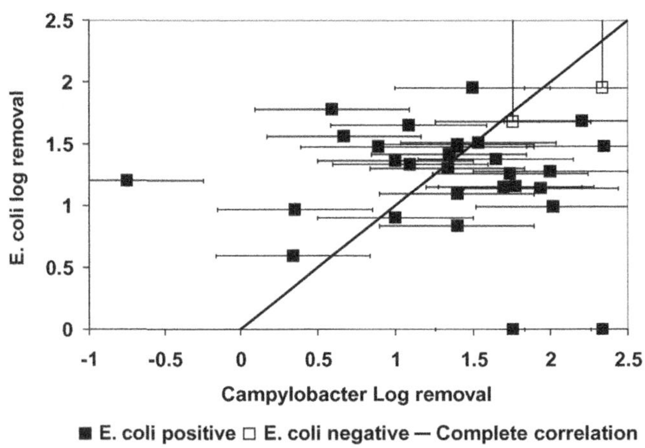


**Figure 6.9** Observed removal of *Campylobacter* and *E. coli* by filtration based on monitoring results paired by date

The reduction by filtration $\pi$ in the stochastic model was calibrated with the *E. coli* data before and after filtration. This calibrated reduction $\pi$ was then applied to monitored *Campylobacter* in the source water in a Monte Carlo simulation. The predicted *Campylobacter* concentration after treatment was compared to monitored concentrations over the same period. Figure 6.10 shows the predicted and monitored *Campylobacter* concentrations. The monitored *Campylobacter* concentrations were within the predicted range. In this case the use of indicator data did provide an accurate model of the level and variability of pathogen reduction by treatment, despite the initially apparent absence of correlation. Direct comparison of *E. coli* and *Campylobacter* reduction by pairing samples on the same date was not valid since the individual observed concentrations were affected by over dispersion of microorganisms in water. However, *E. coli* and *Campylobacter* are similar organisms with respect to their removal by filtration (size and surface characteristics and survival in the environment). Therefore both organisms were removed similarly. Although the momentary removal might not be identical at one moment in time, the level and variability of removal over a longer period was very similar.

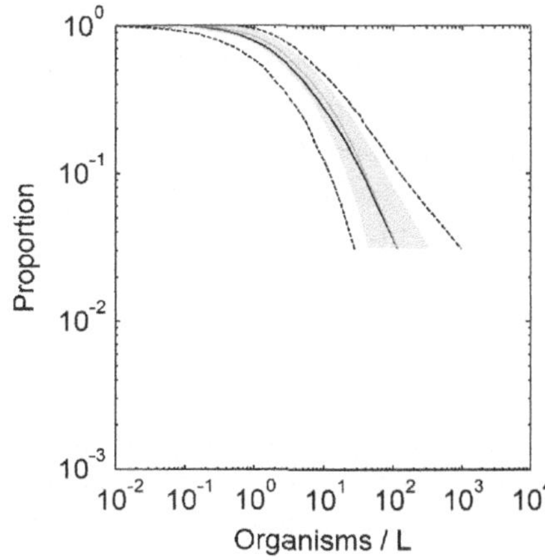


**Figure 6.10** Predicted (lines) and monitored (grey area) *Campylobacter* concentration after filtration. *E. coli* removal was used to model removal of monitored *Campylobacter* concentrations in source water

## DISCUSSION

### Treatment assessment

Microbial monitoring data at full-scale treatment systems provides the most direct information on site specific treatment performance with respect to pathogen reduction. This study showed that different methods to determine treatment efficacy from this data provide different results. Therefore the stochastic treatment model, consisting of Monte Carlo simulation, was used to validate the assessed treatment efficacy. Assessing treatment efficacy $\pi$ could be regarded as calibration of the stochastic model. Using the same data for calibration and validation should result in optimal prediction of effluent concentrations, if $\pi$ was correctly calibrated. The assessments made clear that the extreme concepts of pairing by date, random pairing and pairing by rank did not provide a calibrated model that resulted in optimal prediction of effluent concentrations when validated with the calibration data. The optimised method was developed that actually used validation to determine the optimal parameters of the PDF of $\pi$. Thus the stochastic model could be considered a 'black box' type of model. The calibrated PDF of $\pi$ then became the assessed treatment efficacy. The calibrated $\pi$ provided an evaluation of the level of treatment efficacy and it's variation. This information could be used by risk managers and operators to determine whether the system was achieving the treatment targets. When the assessment indicates that treatment efficacy was highly variable, treatment optimisation could be applied to reduce the probability of rare events of poor treatment.

## Stochastic model calibration

Stochastic modelling of drinking water treatment for QMRA purposes was applied to a range of full-scale microbial monitoring data. Most reported QMRA studies in literature calibrated the stochastic model with the 'date method', where influent and effluent samples were paired by date. The analysis of full-scale data showed that pairing by date generally resulted in the same estimation of $\pi$ as randomly pairing influent and effluent data. This indicated that there was no significant correlation in time between influent and effluent samples in the assessed full-scale treatment processes. The distribution of $\pi$ calibrated with the date method showed a large range, including $\pi$ values $>1$ (treatment generates pathogens). Application of the date method to model the systems in Table 6.1 resulted in overestimation of concentrations after treatment, sometimes by several orders of magnitude. The 'rank method' for calibration of $\pi$ (Smeets *et al.*, 2008) generally provided a more accurate validation of the model, although in some cases the effluent concentrations were underestimated. The optimized method resulted in calibration of $\pi$ that, by definition, resulted in optimal validation with the calibration data. The optimised estimate of $\pi$ was generally identical or close to the rank method estimate. This indicated that sampling variations (due to over dispersion of microorganisms in water) had a significant effect on the observed reduction.

The point estimate of $\pi$ calibration by the mean in/out method concentrations provided a fairly accurate estimate of the mean treatment efficacy $\pi$ in 60% of the assessments, but over predicted $\pi$ in 30% of the assessments. The mean in/out method provided no information on treatment variability. The optimised method should be preferred since it provides a more accurate estimate of mean treatment efficacy and provides information on variability of treatment efficacy based on the same data as the mean in/out method.

## Stochastic model applications

The stochastic model calibrated with the optimised method was put to the test in several applications that could be part of a QMRA study. First the validity and accuracy of model predictions was tested. Using five years of data to calibrate $\pi$, the model was able to predict the yearly distribution of effluent concentrations based on the monitored influent concentrations. The example used a five year period to calibrate the model to reduce sampling error that may have occurred in the yearly observations. Using only one year of data would therefore have resulted in a less accurate model. This example showed that stochastic modelling to predict effluent concentrations was valid when system conditions were similar during calibration and prediction periods.

Predicting pathogen reduction based on indicator monitoring data is a common application of stochastic treatment modelling in QMRA. The example of using *E. coli* removal by filtration to predict *Campylobacter* removal in this study showed that indicator organism data could be used for model calibration of pathogen removal. Still correlation between removal of both organisms was not observed directly when data was paired by date due to the over dispersion of organisms in water. This is similar to the shortcomings of the date method. The use of indicator organisms has several advantages. Microbial methods are generally simpler, cheaper, faster and provide a better quantified result (direct count methods and high recovery) than for pathogens. Indicator monitoring is already part of legal compliance monitoring in most countries, therefore the samples serve a dual purpose and historical data can be used for initial treatment assessments. Given the fact that many water utilities have a wealth of data on indicators in their systems, this data could well be used to rapidly increase our knowledge of drinking water treatment processes. The assessments presented in Table 6.1 provide a first overview of actual full-scale treatment efficacy.

QMRA is often referred to as a 'data hungry' method. This study made clear that sufficient microbial monitoring is essential for a accurate site specific assessment of treatment efficacy. However, it is the variability of the system that requires this amount of monitoring, not the QMRA method. The QMRA method provides insight in the uncertainty of the assessment, which can be reduced by additional monitoring. Alternative methods, such as log credits do not account for site specific differences and provide no indication of the uncertainty of the assessment. This results in a false sense of accuracy (and possibly safety) of log credit methods.

## CONCLUSIONS

QMRA is increasingly used for decision support, e.g. in the water safety plan, or even for legal compliance. Stochastic modelling of pathogen reduction by treatment is an effective way to estimate pathogen concentrations in drinking water. Microbial monitoring before and after treatment steps provides the most direct site specific information on full-scale treatment performance. This study showed that the current method to calibrate stochastic models by pairing monitoring results of influent and effluent by date was inaccurate. Treatment efficacy was underestimated resulting in overestimation of pathogen concentrations in the effluent, sometimes by several orders of magnitude. An alternative method presented in this paper resulted in optimal prediction of effluent concentrations. The calibrated model provided an assessment of the level and variability of pathogen reduction by treatment. The model was effectively applied to case studies to predict pathogen concentrations in the effluent based on monitored influent concentrations. The assessment also showed that indicator organism data can be used to calibrate the stochastic model for pathogen reduction. The methods presented in this study are relatively simple to implement by risk managers and results can effectively and intuitively be evaluated visually. Simple methods for treatment quantification in QMRA, such as log credits, can provide a false sense of accuracy and security and do not take site specific factors into account.

## REFERENCES


Gerba C. P. and Haas C. N. (1988). Assessment of risks associated with enteric viruses in contaminated drinking water. *ASTM Special Technical Publication*, **976**, 489–494.

Gibson III C. J., Haas C. N. and Rose J. B. (1999). Risk assessment of waterborne protozoa: current status and future trends. *Parasitology*, **117**, 205–212.

Haas C. N. (1983). Estimation of risk due to low doses of microorganisms: A comparison of alternative methodologies. *American Journal of Epidemiology*, **118**, 573–82.

Haas C. N., Rose J. B., Gerba C. P. and Regli S. (1993). Risk assessment of virus in drinking water. *Risk Analysis*, **13**(5), 545–552.

Haas C. N. and Trussel R. R. (1998). Frameworks for assessing reliability of multiple, independent barriers in potable water reuse. *Water Science and Technology*, **38**(6), 1–8.

Haas C. N., Rose J. B. and Gerba C. P. (1999). *Quantitative Microbial Risk Assessment*. Wiley, New York, USA.

Hijnen W. A. M., Schijven J. F., Bonné P., Visser A. and Medema G. J. (2004). Elimination of viruses, bacteria and protozoan oocysts by slow sand filtration. *Water Science and Technology*, **50**, 147–154.

Hijnen W. A. M., Beerendonk E. and Medema G. J. (2005a). *Elimination of microorganisms by drinking water treatment processes (second edition)*. Kiwa NV, Nieuwegein, the Netherlands.

Hijnen W. A. M., Beerendonk E. and Medema G. J. (2005b). Additional Memo Including Rapid Sand Filtration, GAC Filtration and Direct Filtration to *Elimination of Microorganisms by Drinking Water Treatment Processes* (Not published). Kiwa NV, Nieuwegein, the Netherlands.

ILSI Risk Science Institute Pathogen Risk Assessment Working Group (1996). A conceptual framework to assess the risks of human disease following exposure to pathogens. *Risk Analysis*, **16**(6), 841–848.

LeChevallier M. and Au K. K. (2004). *Water Treatment and Pathogen Control Process: Efficiency in Achieving Safe Drinking Water*. WHO, IWA Publishing, London, UK.

Masago Y., Katayama H., Hashimoto A., Hirata T. and Ohgaki S. (2004). Assessment of risk of infection due to *Cryptosporidium parvum* in drinking water *Water Science and Technology*, **46**(11–12), 319–324.

Medema G. J., Groennou J., Hijnen W. A. M., Teunis P., Kruidenier L., Ketelaars H., Willemsen-Zwaagstra J., Havelaar A. and Van der Kooij D. (1999). Frequency Distributions of *Cryptosporidium* and *Giardia* in Raw Water and Elimination Capacity of Water Treatment. In: *Cryptosporidium and Giardia: New Challenges to the Water Industry*, Medema G. J. Thesis, University of Utrecht, the Netherlands.

Payment P., Berte A., Prévost M., Ménard B. and Barbeau B. (2000). Occurrence of pathogenic microorganisms in the Saint Lawrence River (Canada) and comparison of health risks for populations using it as their source of drinking water. *Canadian Journal of Microbiology*, **46**, 565–576.

Regli S., Rose J. B., Haas C. N. and Gerba C. P. (1991). Modeling the risk from *Giardia* and viruses in drinking water. *Journal of the American Water Works Association JAWWA*, **83**(11), 76–84.

Rose J. B., Haas C. N. and Regli S. (1991). Risk assessment and the control of waterborne *Giardia*sis. *American Journal of Public Health*, **81**(6), 709–713.

Smeets P. W. M. H., Dullemont Y. J., Van Gelder P., Van Dijk J. C. and Medema G. J. (2008). Improved methods for modelling drinking water treatment in quantitative microbial risk assessment: A case study of *Campylobacter* reduction by filtration and ozonation. *Journal of Water and Health*, **6**(3), 301–314.

Teunis P. F. M., Havelaar A. H. and Medema G. J. (1994). *A literature Survey on the Assessment of Microbiological Risk for Drinking Water*. RIVM Rapport 734301006, RIVM Bilthoven, the Netherlands.

Teunis P. F. M., Medema G. J., Kruidenier L. and Havelaar A. H. (1997). Assessment of the risk of infection by *Cryptosporidium* or *Giardia* in drinking water from a surface water source. *Water Research*, **31**(6), 1333–1346.

Teunis P. F. M. and Havelaar A. H. (1999). Cryptosporidium *in drinking water: Evaluation of the ILSI/RSI Quantitative Risk Assessment Framework*. Report No. 284550006 RIVM, Bilthoven, the Netherlands.

Teunis P. F. M, Evers E. G. and Slob W. (1999). Analysis of variable fractions resulting from microbial counts. *Quantitative Microbiology*, **1**, 63–88.

Teunis P., Davison A. and Deere D. (2004). *Short-term Fluctuations in Drinking Water Quality and their Significance for Public Health*. Report, World health organization, Geneva, Switzerland.

Westrell T., Bergstedt O., Stenström T. A. and Ashbolt N. J. (2003). A theoretical approach to assess microbial risks due to failures in drinking water systems. *International Journal of Environmental Health Research*, **13**(2), 181–197.

# Chapter 7

# Practical applications of quantitative microbial risk assessment for water safety plans

*P.W.M.H. Smeets[a,b], L.C. Rietveld[b], J.C. van Dijk[b] and G.J. Medema[a,b]*

*[a]KWR Watercycle research institute, PO Box 1072, 3430 BB Nieuwegein, the Netherlands*
*[b]Faculty of Civil Engineering, Delft University of Technology, PO Box 5048, 2600 GA Delft, the Netherlands*

## ABSTRACT

The absence of indicator organisms in drinking water does not provide sufficient guarantee for microbial safety. Therefore the water utilities are implementing water safety plans (WSP) to safeguard drinking water quality. Quantitative microbial risk assessment (QMRA) can be used to provide objective quantitative input for the WSP. This study presents several applications of treatment modelling in QMRA to answer the risk managers questions raised in the WSP. QMRA can estimate how safe the water is, how much the safety varies and how certain the estimate of safety is. This can be used in the WSP system assessment to determine whether treatment is meeting health-based targets with the required level of certainty. The QMRA methods use site specific full-scale information to quantify and reduce uncertainties caused by log-credit assessments that are currently applied in the WSP. QMRA also provides decision support on other issues of the WSP such as designing physical and microbial monitoring, setting critical limits, optimising treatment and preparing corrective actions. Thus QMRA can contribute to efficient and effective management of microbial drinking water safety.

## INTRODUCTION

At the start of the twentieth century, the use of coliforms as indicator organisms to judge the microbial safety of drinking water was initiated (Greenwood and Yule, 1917). Verification of the absence of indicators in treated

 *Stochastic Modelling of Drinking Water Treatment in Quantitative Microbial Risk Assessment.* By Patrick Smeets. ISBN: 9781843393740. Published by IWA Publishing, London, UK.

water samples is still part of most legislation today. The applicability of the indicator concept turned out to be limited, as outbreaks of infectious disease continued to occur (Hrudey and Hrudey, 2004). By 1990 two developments to assess and improve the microbial safety of drinking water were started; Water Safety Plans (WSP) and Quantitative Microbial Risk Assessment (QMRA).

The development of the WSP concept started in 1994 when the use of Hazard Analysis and Critical Control Point (HACCP), as applied for food safety, was tested for applicability in drinking water safety (Havelaar, 1994 and Teunis *et al.*, 1994). Over the years this concept developed into WSP (Barry *et al.*, 1998; Deere and Davison, 1998). The use of the WSP to manage drinking water safety in an integrated manner was promoted by the Bonn charter (IWA/WHO, 2004) and the third edition of the Drinking Water Guidelines (WHO, 2004).

QMRA development started after the implementation of chemical risk assessment for the Safe drinking water act (SDWA, 1974). Between 1983 and 1991 risk assessment was used sporadically to assess microbial risks in drinking water (Haas, 1983; Gerba and Haas, 1988; Rose *et al.*, 1991 and Regli *et al.*, 1991). In 1996 the ILSI Risk Science Institute Pathogen Risk Assessment Working Group developed a conceptual framework to assess the risks of human disease associated with exposure to pathogenic microorganisms (ILSI, 1996). The framework was later evaluated by Teunis and Havelaar (1999) and more or less resulted in an extensive guide to risk assessment for pathogens in (drinking) water (Haas *et al.*, 1999). In the USA, the early QMRA studies formed the basis for the Surface Water Treatment Rule (SWTR, USEPA, 1989). Technical requirements for drinking water treatment were defined in the SWTR guided by a risk target of $10^{-4}$ probability of infection per person per year for *Giardia* and viruses. Later the SWTR was extended for *Cryptosporidium* in the IESWRT (USEPA, 1998), and was further elaborated in the LT1ESWTR (USEPA, 2002) and LT2ESWTR (USEPA, 2006). Published QMRA studies generally concluded that quantifying treatment efficacy currently introduces the most uncertainty (Teunis *et al.*, 1997; Gibson *et al.*, 1999 and Payment *et al.*, 2000). QMRA studies have used different types of risk endpoints such as the yearly average risk of infection (Anonymous, 2001), daily risk of infection Signor *et al.* (2007) and disability adjusted life years (DALY) (WHO, 2004). The endpoint of yearly average risk of infection was used in this study, since it allowed effective demonstration of QMRA applications for WSP without introducing the complicating factors of the DALY.

The goal of the WSP is to manage water supply such that health-based targets are met (Davison *et al.*, 2006). To determine if the health-based targets are met, the risk of infection from the supplied water needs to be assessed. Qualitative and semi-quantitative estimates of risk have been applied in WSP, such as the risk matrix (Davison, 2006). The risk matrix relies on the previous experiences of those involved in the WSP development. The risk matrix requires detailed quantitative judgement on the effect of a risk such as 'compliance impact (minor)' or 'public health impact (catastrophic)'. This method leads to a subjective judgement of risk that is suitable to identify 'clear' risks. A 'tier one' level of risk assessment is generally applied, using log credits and CT tables to quantify treatment efficacy (Davison *et al.*, 2006). Log credits could be effective at a screening level QMRA to prioritize between different sites and to identify sites that are likely not to achieve the health-based targets (Medema *et al.*, 2006). However, log credits do not account for site specific conditions and variations of treatment efficacy in time and the uncertainty involved (Teunis *et al.*, 2004). The assessment of risk is relevant for several elements of the water safety plan. Chapter 1 discussed the relevance of risk estimation for several parts of the WSP:

- verifying compliance with health-based targets
- identify risk events
- prioritize risks
- design monitoring
- setting critical limits
- define corrective actions

The goal of the study was to improve risk estimation in the WSP by including variability and uncertainty in the estimate using stochastic QMRA methods, and to demonstrate how this could provide decision support for the aforementioned parts of the WSP for which risk estimation is relevant.

## METHODS

The method of QMRA has been developed up to the point where a more or less standard framework is applied (Haas *et al.*, 1999). Smeets *et al.* (2008) improved these methods for the analysis of full-scale microbial monitoring data. The risk for a specific system was assessed by identifying the pathogen sources, monitoring pathogens in source water, describing treatment performance, contamination during distribution and consumption and applying dose-response relations. Since all these elements vary in time and space, each was described by a Probability Density Function (PDF). The PDFs for each element were combined in a Monte Carlo simulation to calculate the risk of infection. This study focussed on quantifying treatment efficacy. To determine the efficacy of a treatment process, the Monte Carlo simulation was performed using the PDFs that described the treatment. A value was repeatedly drawn randomly from each of the PDFs and treatment efficacy was calculated for each combination.

Data on treatment from 12 catchment to tap systems (CTS), was collected in the MicroRisk project which represent the current status of treatment monitoring. This data included a description of the system, microbial monitoring data of raw water and treated water at different stages in treatment, non-microbial (on-line) monitoring data on treatment performance such as turbidity and chlorine residual, and process conditions such as flow rates and temperatures. From this data, suitable examples were selected to illustrate the developed methods. Ozonation at CTS1 and chlorination at CTS10 were studied in detail. The ozonation system at CTS1 consisted of two parallel contactors of 167 $m^3$ with no baffling, the average contact time was 6 minutes. Chlorination at CTS10 took place in a single unbaffled contact chamber with an average contact time of 40 minutes. Disinfection was modelled assuming Chick-Watson disinfection kinetics (Chick, 1908) and a Continuously Stirred Tank Reactor (CSTR) in series model (USEPA, 2006). Application of these models and applicable inactivation parameters were described in Smeets *et al.* (2005, 2006). A literature review was used to acquire nominal performance data for physical treatment processes (Hijnen, 2005a, 2005b). Calculations were performed in Microsoft Excel and in Matlab® (including Statistics toolbox). Reduction of pathogens was calculated as

$$\pi = \frac{N_{out}}{N_{in}} \tag{1}$$

Where $\pi$ is the reduction of pathogens, $N_{in}$ and $N_{out}$ (organisms/L) are the pathogen densities before and after treatment respectively. Log reduction or Decimal Elimination (DE) was calculated as $\log_{10}(\pi)$.

A loss of treatment efficacy due to a special event, e.g. a clogged dosing pump, was referred to as a failure. The acceptable risk criteria for failures were expressed in the form of Equation 2 where $\alpha_{acc}$ is the acceptable contribution of failures to average risk, $\pi_n$ is the nominal treatment efficacy, $\pi_f$ is the treatment efficacy during failure and $p_f$ is the proportion of time that failure occurs.

$$\alpha_{acc}\pi_n \geq \pi_f p_f \tag{2}$$

The probability of detecting a failure event was calculated with Equation 3. Assuming that monitoring indicated either 'non-failure' or 'failure' of the process, the chance of detecting a failure event $p_d$ could be determined from the chance of failure $p_f$ and the number of samples $N$:

$$p_d = 1 - (1 - p_f)^N \tag{3}$$

The mean treatment efficacy including failures $\pi_m$ was calculated as:

$$\pi_m = (1 - p_f)\pi_n + p_f \pi_f \tag{4}$$

The average yearly treatment efficacy including corrective actions was calculated with Equation 5 which is similar to Equation 4 but included three situations: nominal treatment, failure and corrective treatment.

$$\pi_m = (1 - p_f - p_c)\pi_n + p_f \pi_f + p_c \pi_c \tag{5}$$

Where $p_c$ was the proportion of time that corrective treatment was performed and $\pi_c$ was the treatment efficacy during corrective treatment.

The following applications of stochastic QMRA for parts of the WSP for which risk estimation is relevant were discussed:

Compliance with health-based targets

- Including uncertainty of log credits
- Reducing the uncertainty of log credits with site specific information
- Including uncertainty of disinfection modelling
- Including site specific variability in disinfection modelling
- Modelling improvements of disinfection processes

Verifying treatment efficacy

- Design of microbial monitoring
- Design of process monitoring

Operation of treatment

- Setting critical limits
- Preparing corrective actions.

# RESULTS

## Compliance with health-based targets

### Including uncertainty of log credits

Risk managers that start to assess their system can account for the uncertainty caused by the lack of site specific information. The following example assesses the uncertainty included when assessing the *Cryptosporidium* removal efficacy of rapid sand filtration. Log credits provide an initial estimate of treatment efficacy (USEPA, 2006), and numerous studies have indicated the variability of treatment (LeChevallier and Au, 2004 and Hijnen *et al.*, 2005a, 2005b). This variation between studies was considered as the uncertainty with respect to the site specific efficacy. Figure 7.1 shows the reported removals in literature as a histogram (Hijnen *et al.*, 2005b). Observed efficacy of *Cryptosporidium* removal by filtration varied from 1.3 to 5.3 log removal. Rather than selecting a conservative or mean log removal value as a point estimate, the whole range of observed removal was described by a triangular probability density function (PDF).

Since the PDF was fit over the log removals, it was actually a log-triangular PDF, therefore the log of the mean reduction did not equal the mean of the log reduction. In the Monte Carlo simulation, the log of the mean reduction represented the expected value of the log-triangular distribution (mean out/mean in). The distribution in Figure 7.1 resulted in an expected value of 1.5 log removal although the mode of the triangular distribution was 2 log removal. The 95% confidence limit (CL) was only 0.8 log removal, so there was a 5% probability that removal was even lower. Given Figure 7.1 there is a high probability that the removal exceeded 2 logs. Therefore collecting more site specific information on treatment efficacy could result in assessment of higher removal at an acceptable CL.

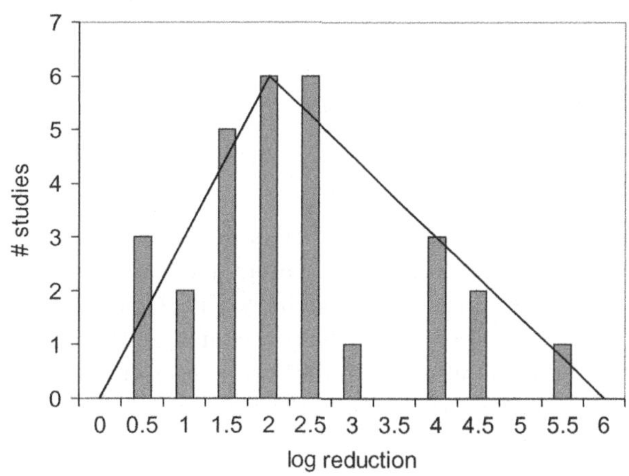


**Figure 7.1** Histogram of log removal of *Cryptosporidium* by filtration reported in scientific literature and the triangular PDF describing uncertainty for stochastic modelling

This information can support decisions by the risk manager. When it is sufficiently likely that the health-based target is met with the expected removal, no specific action is required. When it is unlikely that the health target is met, the risk manager can decide to adapt the system (e.g. additional treatment) without requiring additional research. When it is uncertain if the target is met, the risk manager can collect more site specific information on treatment efficacy to demonstrate that low efficacy does not occur at this specific site. Collecting more information could be substantially more cost efficient than adapting treatment.

## Reducing uncertainty of log credits with site specific information

A major source of uncertainty when applying log credits is the site specific performance of a treatment system. Site specific data can be used to reduce this uncertainty, for example through the use of surrogate information such as turbidity after filtration and turbidity removal. Studies have shown that turbidity removal is not a direct index of pathogen removal, yet a good working filter is able to produce water with a turbidity constantly below 0.1 NTU (Hijnen, 2005a, 2005b). This knowledge was used to refine the triangular PDF of nominal removal that was applied to the log credits. Several CTSs in the MicroRisk project monitored reduction of indicator organisms and turbidity by conventional treatment processes. The overview of the results in Table 7.1 shows that indeed the reduction of bacteria was highest at filtration sites with very low turbidity after filtration and substantial turbidity reduction. At CTS 11 high turbidity values coincided on a daily basis with positive bacterial results, whereas at CTS 2 no clear relation was found (daily data not shown).

**Table 7.1** Observed relation between effluent turbidity, turbidity removal and treatment efficacy for removal of *E. coli* and spores of sulphite-reducing Clostridia (SSRC)

| CTS | Process | Effluent turbidity | Log reduction | | |
|---|---|---|---|---|---|
| | | | Turbidity | *E. coli* | SSRC |
| 1 | Filtr | 0.1 | 1 | 0.5 | 0.2 |
| 1 | coag-sed-filtr | 0.1 | 2 | 1.4 | 1.8 |
| 2 | Filtr | 0.2 | 0.9 | 1.3 | 1.3 |
| 5 | coag-sed-GAC | $<0.05$ | 2 | $>3$ | – |
| 6 | coag-sed-GAC | $<0.05$ | 0.9 | $>1$ | – |
| 9 | Filtr | 0.1 | 0.8 | 1 | 1.6 |
| 10 | coag-sed-O3-filtr | 0.4 | 1.2 | – | 1.2 |
| 11 | direct filtr | 0.05 | 2 | 2.7 | – |

Two examples were used to demonstrate how log credits could be adapted using site specific data. CTS 10 provided an example where turbidity was measured before and after conventional treatment (coagulation-sedimentation and filtration). The filtrate turbidity varied between 0.1 and 1 NTU, indicating that filtration did not work effectively, so a PDF for a poorly working filter was applied (Figure 7.2). Turbidity at CTS 11 was recorded daily before and after direct filtration. Turbidity was consistently reduced from $>1$ NTU to $<0.06$ NTU thus verifying that the filter was working well, so the triangular PDF for a well performing filter was applied. Figure 7.2 illustrates how the triangular PDF was adapted for these CTSs based on the recorded turbidity. The maximum of the PDF for poor performance was set to the original mode of the PDF, assuming a symmetrical triangular distribution. Similarly, the minimum of the PDF for good performance was set to the original mode of the PDF, again assuming a symmetrical triangular distribution. The choice of level of adaptation of the PDF was rather arbitrary in this study. However, the approach did provide a tool for risk managers to differentiate between systems based on site specific information.

## Including uncertainty of disinfection modelling

Deterministic models generally describe disinfection as a first order reaction, where the achieved inactivation depends on disinfectant concentration $C$, contact time $t$ and the inactivation rate constant for the microorganism $k$. Process conditions such as $C$ and $t$ vary in full-scale systems and may be hard to determinate exactly. The inactivation rate constant varies with temperature and values for specific disinfectants and pathogens reported in literature vary between studies (USEPA, 2006 and Smeets *et al.*, 2005, 2006). These uncertainties and variations were included in a stochastic disinfection model in a case study of *Campylobacter* inactivation by ozonation at CTS1 by considering all the parameters and variables ($C$, $t$ and $k$) as a stochastic parameter. To estimate disinfection with a CSTR disinfection model (USEPA, 2006) the parameters $k$, $t$ and $C$ needed to be determined.

Various ozone inactivation rates found for *E. coli* were applied for *Campylobacter* inactivation (Smeets *et al.*, 2005). Inactivation rates of 100 to 10,000 $L.mg^{-1}.min^{-1}$ were reported within the temperature range of 1 to 30°C. Conservative rates were in the order of 100–430 $L.mg^{-1}.min^{-1}$ and optimistic rates were 2,000–10,000 $L.mg^{-1}.min^{-1}$ depending on temperature. The range of inactivation constants exceeded the effect of temperature, therefore temperature was not modelled separately from the inactivation range. The inactivation rate $k$ was modelled stochastically by uniformly distributing $\log_{10}(k)$ from 2 to 4 in the stochastic model.

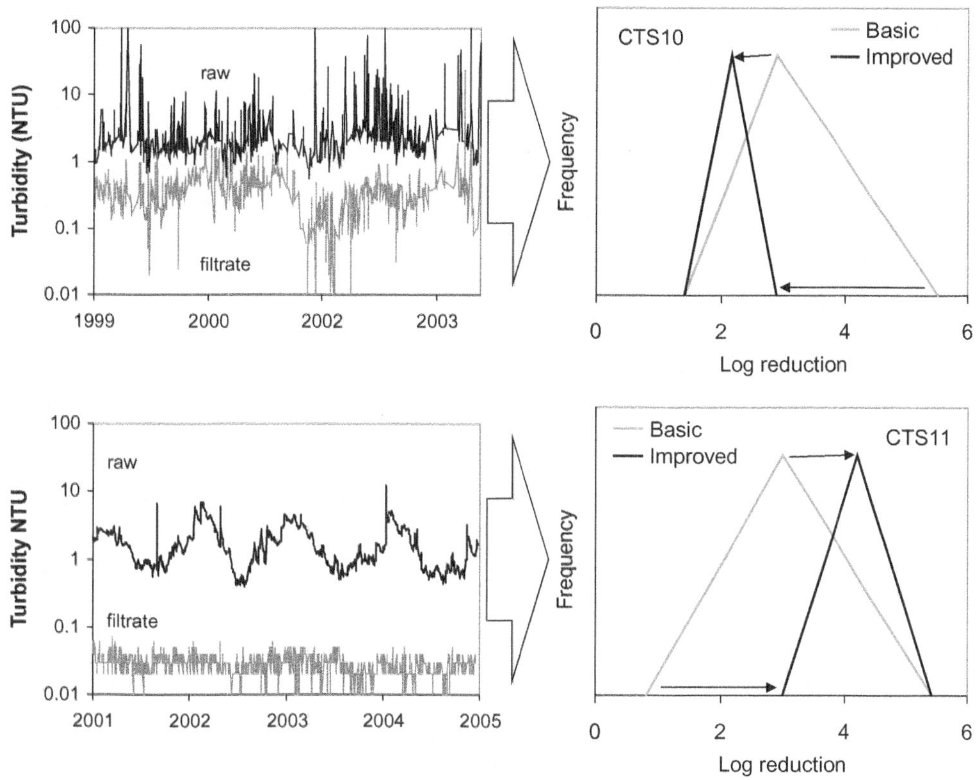


**Figure 7.2** Turbidity before and after conventional treatment (CTS10) and direct filtration (CTS11) was used to refine the triangular PDF for 'good' or 'poor' performing filters

Contact time of each part of the water was one major uncertainty in the modelling of disinfection systems. In this example a continuously stirred tank reactor (CSTR) model was used to account for residence time distribution (USEPA, 2006). A single CSTR reflects the conditions in a single, unbaffled contact chamber, multiple CSTRs in series reflect more plug-flow like conditions. No tracer experiments were available to determine the appropriate number of CSTRs for the case study system. A single CSTR system was the most conservative estimate applied. The most optimistic estimate applied a 3 CSTR model, and the stochastic model applied a uniform discrete distribution of 1 to 3 CSTRs (Do-Quang *et al.*, 2000).

The ozone dose was controlled by monitoring ozone concentrations in the ozone contactor and adjusting the dose to reach the setpoint. The measured ozone concentration only represented a small part of the total water volume. Van der Veer *et al.* (2005) showed that ozone concentrations can vary considerably over a cross-section due to limitation of mixing in ozone bubble columns. Figure 7.3 shows the measured ozone concentrations over the cross section of flow passage from the bubble column to the first contact chamber of a pilot installation. Ozone was measured at each of the intersections of the grid A1, A2 … J5 by a tube sampler. Although site specific differences will influence the pattern, a similar distribution of ozone concentration over the cross section of the water in a full-scale ozonation system could be expected. A normally distributed PDF with a standard deviation of 10% described the variation of the measured concentration relative to the average concentration in Figure 7.3. The same PDF was applied to the ozone concentration at the case study site to reflect the variability of the ozone concentration in the stochastic model. As a result, ozone concentrations $C$ applied in the stochastic model varied from 50% to 150% of mean concentration.

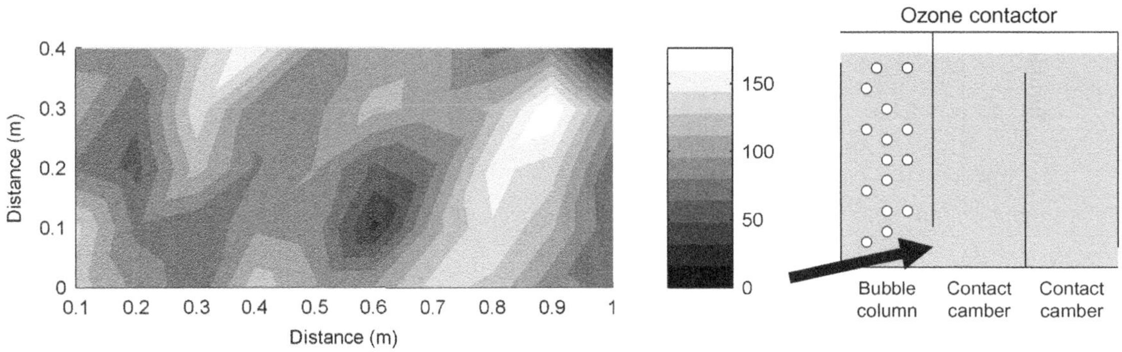


**Figure 7.3** Distribution of measured ozone concentrations (μg/l) over the cross section of the passage from the bubble column to contact chamber of a pilot ozone installation. Ozone was measured at each intersection of the grid A1, A2 … J5. Ozone dose was 1,000 μg/l (Van der Veer *et al.*, 2005)

Three different approaches to include the uncertainty of disinfection modelling were applied to estimate the inactivation of *Campylobacter* based on the mean ozone concentration $C$ and mean contact time $t$. The conservative approach used conservative values for all parameters and variables (1 CSTR, low inactivation rate constant). The optimistic approach used optimistic values (3 CSTRs, high inactivation rate constant). The stochastic approach used the described distributions of the variables and parameters in a Monte Carlo simulation of the treatment system. The mean $C$ column in Table 7.2 shows that choosing conservative or optimistic parameters resulted in 2.7 and 10.8 log reduction respectively. Using a point estimate for the 'worst case' or 'best case' situation would leave the risk manager with 8 orders of magnitude uncertainty about the efficacy of ozonation exceeded. Using a stochastic model the impact of the uncertainties was quantified, resulting in an estimated reduction of 3.5 log and a 95% CL that the inactivation exceeded 2.8 logs. The stochastic model indicated that there was a 50% probability that inactivation exceeded 6 logs. With this information the risk manager could decide whether treatment efficacy was sufficiently verified or whether additional site specific information should be collected to reduce uncertainty or to verify a higher level of inactivation.

**Table 7.2** Estimate of over-all log inactivation of *Campylobacter* by a case study ozonation system using different types of models

| | **Mean *C*** | **Site specific on-line monitoring data** | |
|---|---|---|---|
| | **Log reduction Nominal (95% CL)** | **Log reduction Ex. events (95% CL)** | **Log reduction Incl. events (95% CL)** |
| Conservative point estimate | 2.7 | 2.7 | 2.6 |
| Stochastic | 3.6 (2.8) | 3.5 (3.4) | 3.4 (3.4) |
| Optimistic point estimate | 10.7 | 10.8 | 4.8 |

## Including site specific variability in disinfection modelling

The previous process model example at CTS 1 used general characteristics of the treatment system, such as operational setpoint and mean residence time and temperature to estimate disinfection efficacy. However, conditions vary in time and short events of poor performance can have a significant impact on average risk. Site specific information on this variation, such as on-line ozone and flow monitoring, could provide verification that health-based treatment targets were met. A one year dataset of on-line ozone measurements at an interval of 20 seconds, resulting in over 1.4 million records, were available from the case study site CTS1 (Figure 7.4). In addition, hourly flow and temperature measurements were available. The on-line data was used to model inactivation, again using the conservative, stochastic and optimistic approach. Several events of short periods of no measured ozone were recorded (see Figure 7.4). The impact of these short events on the assessed over-all log reduction over the year depended on the applied method. In one assessment ozone concentrations below 0.1 mg/l were excluded from the dataset to illustrate the effect of events. Table 7.2 provides an overview of the results. The stochastic and conservative approaches were not affected by the use of site specific monitoring data. The optimistic assessment was affected by the occurrence of short moments of ozone failure. These reduced the efficacy over the total period by 6 log units. This example showed that site specific data can reveal short periods of sub-optimal conditions that can affect the mean treatment efficacy, especially when nominal treatment efficacy is very high. Conservative approaches may underestimate treatment efficacy leading to unnecessary actions. Stochastic modelling can provide a realistic estimate of treatment efficacy even when site specific data is limited.

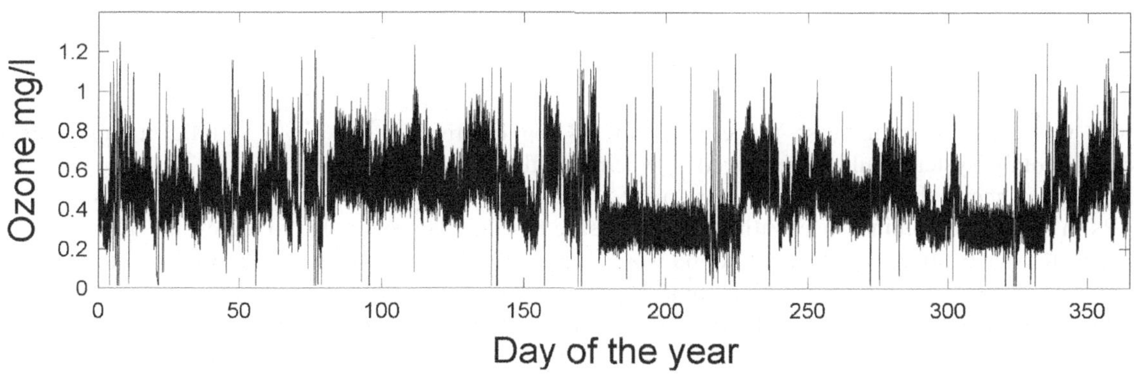


**Figure 7.4** On-line ozone measurements at the case study site at a 20 second interval over a one year period

## Modelling improvements of disinfection processes

When the system assessment in the WSP indicates that health targets are not met, measures are required to improve the system. One of those measures could be to improve an existing disinfection process. This example discusses the chlorine disinfection process at CTS 10. The system assessment showed that over 3 log virus inactivation needed to be achieved by the chlorination to reach health-based targets. Inactivation of viruses by chlorination at CTS10 was modelled with the CSTR model (USEPA, 2006) using monitored chlorine residual and temperature data. Flow was assumed to be constant. The temperature dependant inactivation rate constant for viruses was deduced from the CT tables in the 'Disinfection profiling and benchmarking guidelines' (USEPA, 1999). The contact chamber was not baffled and based on general hydraulic characteristics it was modelled as a single CSTR (DoQuang *et al.*, 2000). Figure 7.5 shows the modelled inactivation achieved over a two month period. Under nominal conditions, 2.5 log inactivation was achieved, but due to the variability of the chlorine level, the mean inactivation over the total period was only 2.1 log.

Several options for improving the efficacy of the process were studied: increasing chlorine dose, improving process control and improving hydraulics. Figure 7.6a shows that doubling the chlorine concentration would increase the inactivation to 2.3 log. This option was therefore ineffective to reach the health target and since it could have negative impact on other goals such as costs and disinfection by-products it was not preferred. A second option was an investment in improved process control (e.g. equipment and software) which would prevent the occurrence of low chlorine levels (Figure 7.6b).

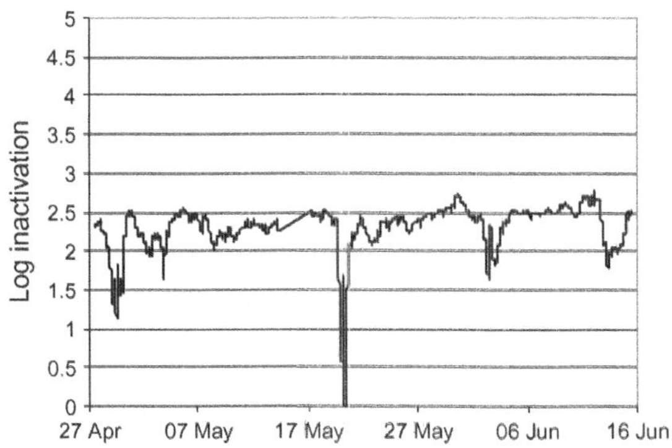


**Figure 7.5** Modelled inactivation of viruses by chlorine at CTS 10 based on on-line monitoring of chlorine residual

This was expected to increase the efficacy to 2.4 log inactivation, so not a significant improvement. Hydraulics of the system could be improved by placing baffles in the contact chamber.

Figure 7.6c shows that if two baffles were placed to improve the hydraulic characteristic of the system to 2 CSTRs in series, the nominal inactivation could increase to 4 log. However, due to the events of low chlorine residuals, the inactivation over the total period would still be limited to 2.6 logs. By improving both hydraulics and process control, 4.1 log inactivation could be achieved (Figure 7.6d). This example shows that a single measure would not be effective and that the right combination of measures would be likely to achieve the required improvements. This type of modelling could be used by risk managers for decision support.

## Verification of treatment efficacy

When the treatment assessment in the WSP indicates that the system is capable of producing safe water, the system needs to be monitored to verify that treatment targets were met. Therefore one of the monitoring goals is to verify that no rare hazardous events have occurred that could have a significant impact on the assessed risk. Since this often requires very frequent monitoring, microbial monitoring is generally not feasible. However, other parameters that may be monitored on-line can be used to verify the system was working as expected. Equipment monitoring such as dosing pump flow can also be used to verify that the system was operational, or to detect moments of failure. Deviation of any of these parameters could lead to reduced treatment efficacy, which may be hard to quantify. A conservative approach is to assume complete failure in case of a deviance. The following examples assume that monitoring either indicates compliance or failure (no pathogen reduction), however if efficacy during failure could be quantified in more detail this could be incorporated in the calculations.

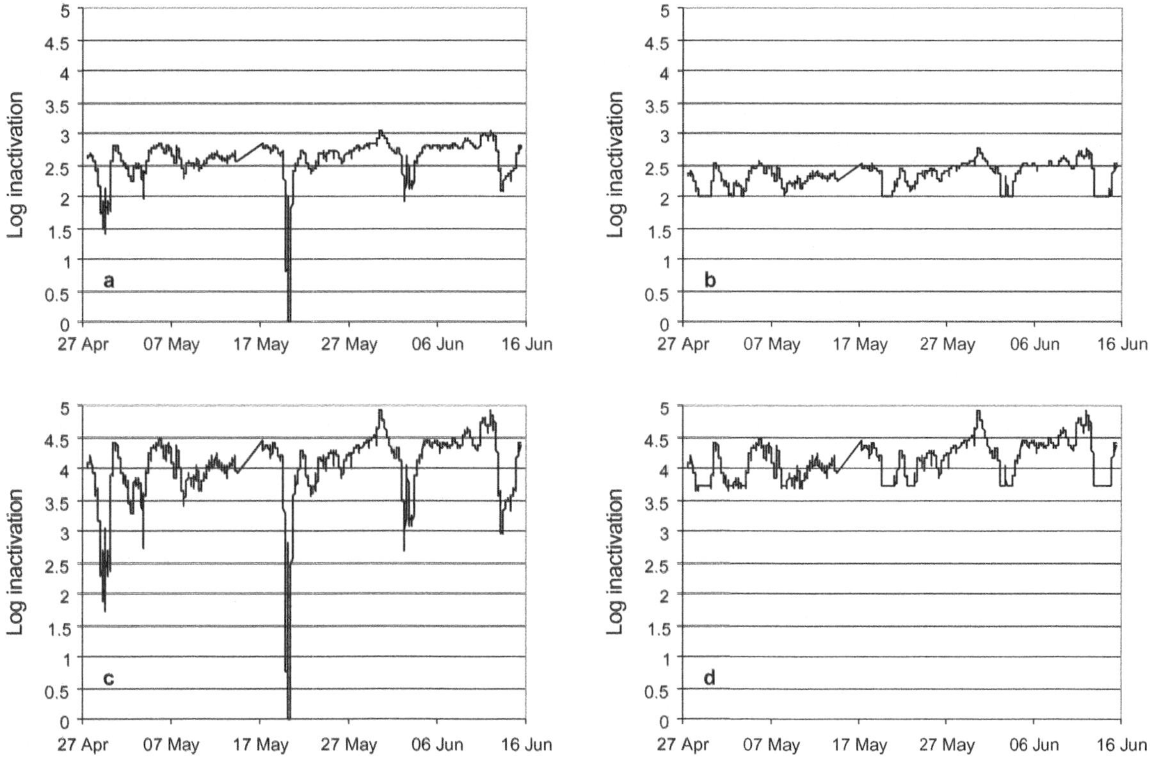


**Figure 7.6** Modelled effect of process improvements on inactivation of viruses by chlorine: doubling chlorine residual concentration (a), improved process control (b), improved hydraulics (c) and improved hydraulics and process control combined (d)

## Design of microbial monitoring

Microbial monitoring before and after a treatment process is the most direct way to assess treatment efficacy. Furthermore microbial monitoring of drinking water could also provide a direct assessment of drinking water safety. The microbial monitoring needs to provide a reliable estimate of the arithmetic mean concentration for both these applications. When organisms in water are over dispersed, high concentrations that rarely occur can still dominate the mean concentration. Monitoring will provide a range of microorganism concentrations, from which the mean concentration is calculated. A question from the risk manager could be: Have I taken enough samples to determine the mean microorganism concentration in the water?

To answer this question, the risk manager needs to determine whether the 'dominant concentrations' have been determined by monitoring. The answer to this question is illustrated through Figures 7.7a, 7.7b and 7.8. The markers show the observed *Cryptosporidium* concentrations in drinking water at three treatment sites in the UK in 2000–2002 (data described in Smeets *et al.*, 2007). The contribution of each monitored concentration to the mean was determined from the proportion of the drinking water with that concentration. A low concentration of 0.01 organism/L occurring 100% of the time resulted in the same mean concentration as 1 organism/L occurring 1% of the time. The dashed line shows combinations of proportion and concentration that contributed equally to the mean concentration. The arithmetic mean concentration at the site in Figure 7.7a was 0.002 oocysts/L. The observed concentrations between $2*10^{-3}$ and $10^{-2}$ oocysts/L dominated this mean concentration. This is where the dashed line touches the markers of the observed concentrations. Based on this data, the average concentration was accurately determined, since higher and lower concentrations would not contribute significantly to the mean concentration. This typical shape was found for 30% of the 216 treatment sites for which data had been collected (Smeets *et al.*, 2007).

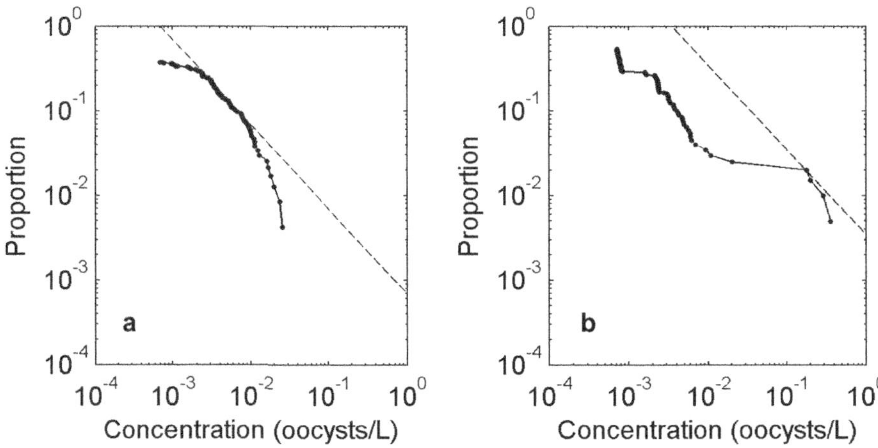


**Figure 7.7** Monitored *Cryptosporidium* in drinking water (markers) including normal events (a) and special events (b). The dashed line indicates combinations of proportion and concentration resulting in the same average concentration

Dominant concentrations generally occurred in less then 1%–5% of the samples in raw water, during treatment and in drinking water. So in order to generate a usable estimate of the risk, over 20 to 100 microbial samples at any point in the system would typically be required.

A complicating factor is the occurrence of 'special events'. Figure 7.7a shows the concentrations due to normal variations, so high concentrations can be considered 'normal events'. Figure 7.7b shows a 'special event' where very high concentrations occurred e.g. due to human error or equipment failure causing a curve break. In Figure 7.7b the dashed line touches the markers at the highest monitored concentration. The concentration during the special event therefore dominated the average concentration and thus on the risk at this site. These special events were observed in 10% of the datasets from 216 treatment sites in the UK. At the example site in Figure 7.7b, other monitoring efforts, apart from microbial monitoring, would be required to verify that these special events did not occur even for a few hours per year (a proportion of $10^{-3}$ translates into eight hours per year). Methods for event monitoring will be discussed later.

A second complicating factor is the detection limit or sample volume. At treatment sites with lower concentrations of *Cryptosporidium* in the treated water, the observed concentrations looked like Figure 7.8 which was typical for 60% of the monitored UK sites. The arithmetic mean concentration was $6*10^{-5}$ oocysts/L based on the monitoring results. This mean concentration was dominated by the lowest observed concentration (at the detection limit) of $7*10^{-4}$ oocysts/L occurring 7% of the time. However, an intuitive interpolation to lower concentrations would exceed the dashed line, thus dominating the mean concentration and therefore the average risk. Larger volumes taken at a lower frequency could be used to determine the dominating concentration. Stochastic modelling of the treatment as performed in QMRA can provide an estimate of these low concentrations, based on raw water concentrations and removal by treatment.

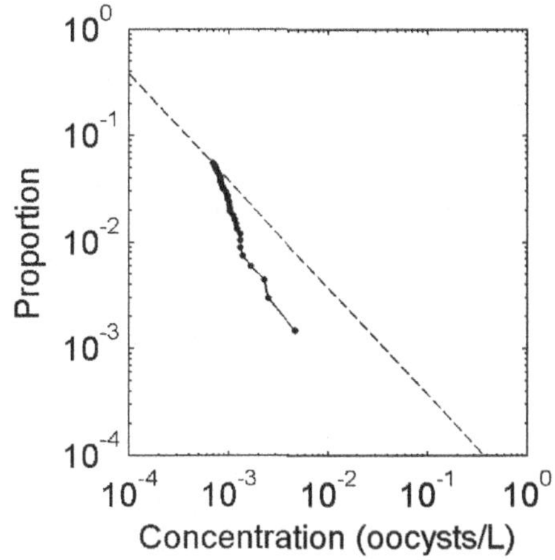


**Figure 7.8** Monitored *Cryptosporidium* in drinking water (markers). The dashed line indicates combinations of proportion and concentration resulting in the same average concentration

The microbial monitoring of drinking water can have three typical outcomes that provide decision support for the risk manager:

- If the dominant concentrations were monitored accurately and therefore the system assessment was accurate (Figure 7.7a);
- If the mean monitored concentration was dominated by the highest observed concentration, additional (more frequent) microbial monitoring is required to determine the dominant concentration (Figure 7.7b);
- If the mean monitored concentration was dominated by the lowest observed concentration, larger sample volumes are required or the concentrations below detection limit need to be estimated through stochastic modelling of the treatment (Figure 7.8).

## Design of process monitoring

First the acceptable level of increase of risk due to failure $\alpha_{acc}$ in Equation 2 needs to be chosen. The risk due to the event will add to the nominal risk. Risks due to failure events might result in a higher average risk than the nominal risk, which would not be considered acceptable. In this example the risk from failures is considered acceptable when on average it equals the nominal risk, so $\alpha_{acc}$ equals 1. A more strict condition could be set by choosing a lower value for $\alpha_{acc}$. Complete failure of the treatment process during an event was assumed in this example, so treatment efficacy during failure $\pi_f$ equalled 1. In that case the acceptable proportion of time that failure occurs $p_f$ equalled nominal reduction $\pi_n$. The probability of detecting a failure event was calculated with Equation 3 assuming that monitoring indicated either 'non-failure' or 'failure' of the process. By choosing the probability of detection $p_d$ at the desired confidence level (e.g. 90%, 95%, 99%) the required number of measurements $N$ could be calculated. The safety of the system would be verified when no failures were detected with $N$ measurements over the assessed period. Table 7.3 shows the required number of measurements in relation to the nominal log reduction to verify at the 95% confidence level that failure events did not significantly increase the risk. The monitoring frequency for the assessment of a one year period was also calculated. Approximately 30% less measurements are required for a 90% confidence level, respectively 50% more measurements are required for a 99% confidence level.

**Table 7.3** Number of required monitoring records to verify at the 95% confidence level that failure events do not significantly add to the risk when compared to nominal reduction

| Nominal log reduction | #/year | Monitoring interval |
|---|---|---|
| 0.05 | 1 | 1 year |
| 1 | 30 | 1 week |
| 2 | 300 | 1 day |
| 3 | 3,000 | 3 hours |
| 4 | 30,000 | 15 min |
| 5 | 300,000 | 2 min |
| 6 | 3,000,000 | 10 sec |
| 7 | 30,000,000 | 1 sec |

Table 7.3 shows that with daily sampling, no more than 2 log reduction can be verified. Above that, more frequent monitoring is required, leading to on-line monitoring. Some treatment plants claim 7 log inactivation of viruses with a disinfection system, so at a 100,000 $m^3$/d plant every 3 litres must be monitored to be 95% confident that all water was sufficiently treated. If the criteria are set more strict, for example setting $\alpha_{acc}$ to 10%, ten times as many samples are required.

A multiple barrier system is easier to monitor. When 6 log inactivation is achieved through 1 barrier (treatment process) 3 million measurements per year are required. The continuous measurement of UV radiation intensity is an example of such intensive monitoring to verifying absence of failure of the highly effective UV disinfection process. When three barriers of 2 logs are placed in series, only 300 measurements per year are required for each process, resulting in 900 measurements in total. So daily monitoring of each process step is then sufficient. This approach assumes that treatment process failures are independent such that failure of the first step does not automatically coincide with failure of the second step. Risk managers can minimise the risk of such consecutive failures in their design of the process for example by separating power supplies and placing physical barriers between equipment such that physical hazards (e.g. flooding or fire) would not affect both processes

The example of the treatment assessment of adjusted log credits at CTS 11 in Figure 7.2 suggested that 3 to 5.5 log reduction could be achieved by this treatment system. Turbidity monitoring could be applied to verify the absence of significant rare events of poor treatment efficacy. According to Table 7.3 a monitoring frequency between 1 and 60 minutes would be required to verify 3 to 5.5 log removal. The excellent turbidity removal in Figure 7.2 suggests that no events of poor removal took place. However, turbidity was only monitored daily, so according to Table 7.3 this was only sufficient to verify 2 log reduction at the 95% confidence limit. Now the risk manager would like to know how certain he could be that a higher level of removal was achieved based on the turbidity data. Given that no failure was observed in 365 records, according to Equation 2 the confidence level for 4 and 3 log reduction was only 4% and 31% respectively. Figure 7.9 shows the confidence levels for the probability that a failure event occurring a certain proportion of time would be detected using daily monitoring.

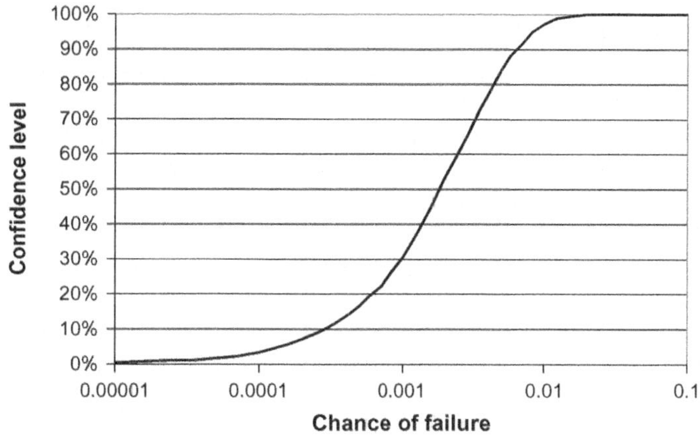


**Figure 7.9** Confidence level of verifying by daily monitoring the absence of a failure in relation to the proportion of time that the failure could occur

At the water treatment plant of CTS 10, the water was disinfected using chlorine. Chlorine residual concentrations were monitored every 15 minutes. Figure 7.10a shows the measured chlorine concentration over two 2.5 month periods. The following example illustrates the impact of monitoring frequency on risk assessment. A subset of the chlorine monitoring results was created to simulate daily sampling (Figure 7.10b).

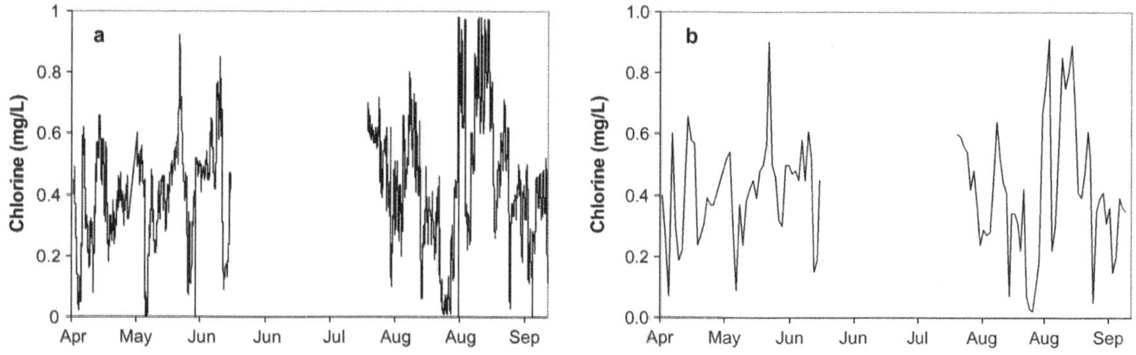


**Figure 7.10** Monitored chlorine concentrations at CTS 10 15 minute data (a) and simulated daily data (b)

Virus inactivation was modelled from these monitored and simulated chlorine measurements assuming constant flow and contact time in a 3 CSTR disinfection model (USEPA, 2006). Virus inactivation rate constants for chlorine at the measured temperatures were deduced from CT tables (USEPA, 1999). Inactivation was calculated for the full dataset of 15 minute samples (10,000 records) and the simulated daily samples (100 records). A Beta distributed probability density function (PDF) was fitted to the disinfection calculated from the selection of daily samples (Figure 7.10b) to extrapolate to rare events of low inactivation. The daily sampling missed several short periods of no chlorine. Therefore, observed inactivation was always more than 2 log and over-all inactivation was 3.8 log. According to Table 7.3 a monitoring frequency of 15 minutes would be required to capture significant rare events. Indeed when the full dataset of 15 minute data was assessed, the short periods of no chlorine were included in the calculations leading to less than 1.5 log reduction 1% of the time and little or no reduction for 0.1% of the time. Over-all inactivation was only 2.3 log. It was very unlikely ($p_d < 10^{-22}$) that increasing the monitoring frequency further would have a significant impact on the assessed log reduction. Theoretically the monitoring frequency could be reduced to verify the 2.3 log inactivation, but it is likely that the water company will strive to improve chlorine control in order to fully use the disinfection potential of 3.8 log, for which the 15 minute interval is required.

Figure 7.11 shows that the fitted PDF did not account for the rare events of no chlorine, so the daily monitoring was insufficient to provide a PDF from which rare events could be extrapolated.

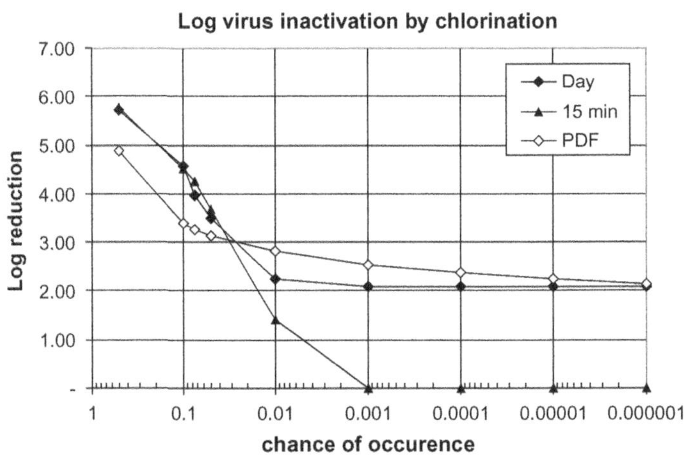


**Figure 7.11** Frequency distribution of calculated log reduction by disinfection based on daily or 15 minute monitoring data and the probability density function (PDF) fitted to the daily data

## Operation of treatment

### Setting critical limits

A treatment system can be designed to provide exactly the right level of treatment to meet the health-based targets. However, in practice the risk manager needs to account for variations and inaccuracies in order to run a practical and stable process. The simulated example in Figure 7.12 is used to illustrate how QMRA can be used to set critical limits and setpoints. The health-based treatment target for the disinfection process simulated in Figure 7.12 was 7.3 log inactivation of viruses. In this example the temperature and flow conditions were constant for the whole period. The required chlorine residual of 1.7 mg/l for 3 log inactivation was determined by disinfection modelling. So if the process was constantly run at exactly this concentration, the yearly health-based target would be met. This level is therefore considered the critical limit. However, in practice the chlorine dose would vary due to operational variation. Chlorine dosing was controlled by an automatic control loop in this example. The chlorine residual was measured and the dosing rate was adjusted if the measured concentration deviated from the setpoint. As a result, the chlorine residual level would typically vary between the operational limits if the system was working within specification, as indicated in Figure 7.12. At hour 20 the chlorine dosing pump got clogged, resulting in a chlorine level below the lower operational limit. This triggered an alarm and the operator was able to clean the pump and restore normal operation. This event did not affect the average treatment performance in a way that the health-based target would not be met, since the critical limit was not exceeded. At hour 80 the chlorine dosing pump failed again, however this time the operator was not in time to restore the system before the critical limit was exceeded. The short period of time that disinfection was ineffective reduced the average treatment efficacy since the critical limit was exceeded. Therefore the operator needed to take corrective actions, such as starting emergency chlorination of the distributed water, in order to achieve the health-based target over the total period.

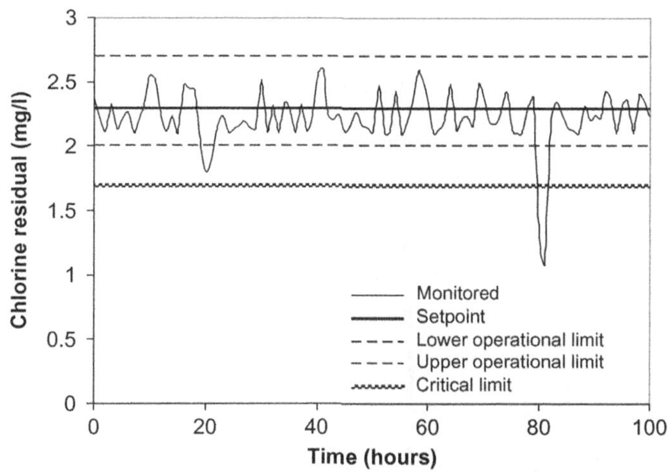


**Figure 7.12** Simulated example of critical limits, operational limits, setpoints and monitoring of a chlorine disinfection process

This example illustrated that setting critical limits was not merely determining the lowest chlorine dose to meet the target, but also included setting operational setpoints and operational limits. Quantifying limits and setpoints is related to many site specific issues such as the treatment target, the normal variability of the process, the probability of events, the response time of the operator and the effect of corrective actions. Simply applying a higher setpoint for chlorine dose can have adverse effects in other objectives (cost, Disinfectant By-Product (DBP) formation, maintenance/operation) and therefore is not an option. QMRA can provide decision support for choosing setpoints and limits when some realistic assumptions are made. The mean treatment efficacy over a period was calculated with Equation 4. The proportion of time that the system fails $p_f$ and treatment efficacy during failure $\pi_f$ can be roughly quantified by means of the risk matrix in the WSP. In the example it was assumed that operator response could take up to 8 hours to restore the system to normal operation, an event could occur once a year and disinfection would fail completely during an event. Given these assumptions, $p_f$ was 8/(365*24), $\pi_f$ equalled 1 and $\pi_m$ equalled 0.001 (3 log reduction). Using Equation 4 the nominal treatment efficacy $\pi_n$ to achieve $\pi_m$ was 0.000085 (4.1 log reduction). Disinfection modelling showed that a chlorine residual of 2 mg/l was required to achieve 4.1 log reduction, therefore the lower operational limit was set to 2 mg/l. Due to normal variation in the process, observed from hour 25 to 75 in Figure 7.12, a setpoint of 2.3 mg/l was needed to maintain this limit. This example illustrated the basic approach to setting limits and setpoints. The method could be adapted to include other variables such as temperature and flow variations and partial treatment failure.

One of the objectives of the multiple barrier system is to balance out peaks or failure at one point by adequate treatment at another point. Total pathogen reduction at a surface water treatment site is often many orders of magnitude due to a large number of treatment steps. The total treatment can be optimized such that the combination of reduction by individual treatment steps combined provides exactly the required reduction to meet the health-based targets. This is illustrated by the circle in Figure 7.13 for the combination of two processes. However, the previous example in Figure 7.12 showed that treatment targets generally need to be set higher to account for temporary failure. This is illustrated by the square in Figure 7.13. Therefore (partial) failure of one step would not directly lead to non compliance in a multiple barrier system.

This example could be extended to an on-line control tool for pathogen reduction to respond to changes in process conditions. The disinfection model could be used in an algorithm programmed into a PLC or SCADA system to constantly adjust the critical limits, operational limits and setpoint for chlorine dosing to maintain the target level of inactivation. The treatment would thus be controlled by the combined effect of process conditions rather than control of individual parameters.

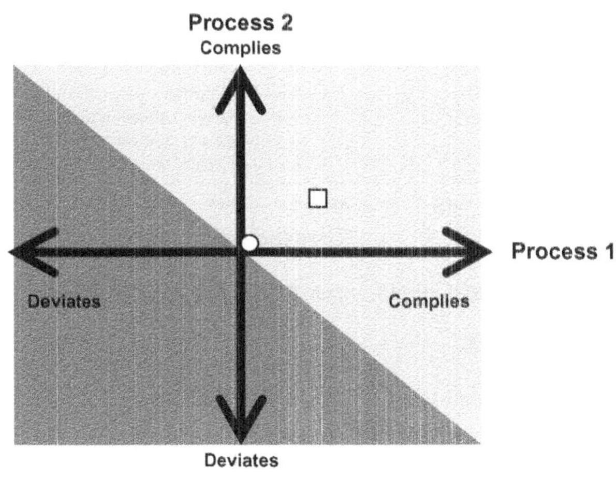


**Figure 7.13** Visualization of combined critical limits in a multiple barrier system. The light grey area illustrates compliance of the total system, the dark grey area illustrates non-compliance

All treatment steps were considered to be independent in this study. In practice some interaction between treatment steps can be expected. Positive interaction occurs when failure at one step leads to failure of a consecutive step(s). For example, failure of pre-oxidation leads directly to less inactivation, but also increases oxidant demand at a later disinfection step, thereby reducing its efficacy. Negative interaction can occur when failure at an early step leads to increased performance of a consecutive step, e.g. poor sedimentation increases particle load to the filters which then work more effectively. Although these mechanisms have been observed, current knowledge is insufficient to create a deterministic model that describes these relations.

## Preparing corrective actions

Even at a well managed system failures could occur that compromise drinking water safety. Such events could be detected by monitoring. A temporal decrease of drinking water quality could have a significant effect on the yearly average risk. When critical limits are exceeded, corrective actions need to bring back water quality to an acceptable level to comply with the yearly risk of infection. This is illustrated with the example of chlorine inactivation of viruses in Figure 7.14. The target inactivation by this system was 2.5 logs, which was generally achieved through process control. The risk manager would need to prepare a plan for corrective actions if failure occurred. Shutting down treatment completely in case of failure was not an option since this would disrupt the other processes and the distribution system would loose pressure thus risking ingress of contaminated water into the distribution system. An emergency UV disinfection unit which could achieve 4.5 log inactivation of viruses could be considered as a corrective measure. The first question for the risk manager was how quickly the emergency equipment needed to be in place. Or, stated differently: how long could this failure be allowed to continue without compromising the treatment target? QMRA could be used to answer this question. Figure 7.14 shows the calculated virus inactivation based on monitored chlorine residual concentration. During normal operation from 27 April to 19 May, generally 2.5 to 3 log inactivation was achieved (black line), resulting in a running average inactivation that complies with the target of 2.5 log inactivation. On 20 May the chlorine dosing failed completely (as an example), resulting in no disinfection. The grey line shows the level of inactivation that could be achieved with the emergency equipment. The average yearly treatment efficacy was calculated for different response times with Equation 5. In this example $\pi_m$ and $\pi_n$ were 0.0032 (2.5 log reduction), $\pi_c$ was 0.000032 (4.5 log) $p_n$ was 139/365 (days until 20 May), $p_c$ was $1 - p_n - p_f$ and $p_f$ varied between 0 hours and 27 days. The dashed line in Figure 7.14 shows the achievable yearly average inactivation in relation to the time the corrective action is started. It was clear from Figure 7.14 that the achievable yearly average quickly decreased in time. In this example, corrective measures needed to be taken within six and a half hours in order to comply with the yearly target of 2.5 log reduction. Risk managers can use similar calculations for decision support on emergency measures and emergency procedures in relation to the health-based treatment target, the critical limits of the treatment process and the achievable corrective action.

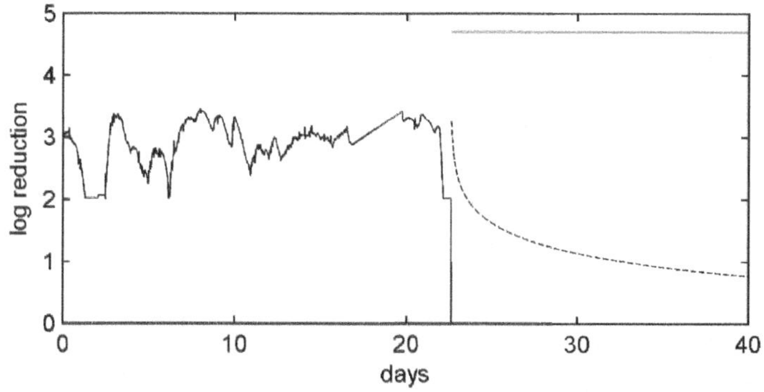


**Figure 7.14** Virus inactivation by chlorine based on chlorine residual monitoring (black line), the level of inactivation during corrective actions (grey line) and the yearly average inactivation that could be achieved by corrective actions in relation to the time required to start corrective actions (dashed line)

## CONCLUSIONS

Water utilities have started to use water safety plans (WSP) to assess and improve the safety of the produced drinking water. Resources for assessments and improvements are limited and therefore need to be used effectively and efficiently where they provide the most benefit for health. Quantitative microbial risk assessment can be used to quantify several questions that are raised in the WSP. Without QMRA the system assessment in the WSP relies on risk manager experience and log-credit from industry standards to quantify drinking water safety. However, this study showed that the resulting uncertainty of the assessed risk can be many orders of magnitude. This uncertainty often makes the difference between compliance or non-compliance with health-based targets. QMRA does not only tell us how safe the water is, but also how much the safety varies and how certain we are that we are meeting health-based targets. The QMRA methods presented in this study can be used to quantify and reduce uncertainties by using site specific full-scale information. QMRA cannot only be used in the system assessment, but also provides decision support on other issues of the WSP. This study provided examples of QMRA to design physical and microbial monitoring, to set critical limits, to support decisions on treatment optimisation and to prepare corrective actions. Thus QMRA can contribute to efficient and effective management of microbial drinking water safety.

## REFERENCES

Anonymous (2001). Adaptation of Dutch drinking water legislation. Besluit van 9 januari 2001 tot wijziging van het waterleidingbesluit in verband met de richtlijn betreffende de kwaliteit van voor menselijke consumptie bestemd water. *Staatsblad van het Koninkrijk der Nederlanden*, **31**, 1–53.

Barry S. J., Atwill E. R., Tate K. W., Koopman T. S., Cullor J. and Huff T. (1998). Developing and Implementing a HACCP-Based Programme to Control *Cryptosporidium* and Other Waterborne Pathogens in the Alameda Creek Watershed: Case Study. *American Water Works Association Annual Conference, 21–25 June 1998, Dallas, Texas Water Resources*, **B**, 57–69.

Chick H. (1908). An investigation of the laws of disinfection. *Journal of Hygiene, Cambridge*, **8**, 92–158.

Davison A., Deere D., Stevens M., Howard G. and Bartram J. (2006). *Water Safety Plan Manual.* World health organization, Geneva, Switzerland.

Deere D. and Davison A. (1998). Safe water – are food guidelines the answer? *Water*, **25**, 21–24.

Do-Quang Z., Ramirez C. C. and Roustan M. (2000). Influence of Geometrical Characteristics and Operating Conditions on the Effectiveness of Ozone Contacting in Fine-Bubbles Conventional Diffusion Reactors. *Ozone: Science and Engineering*, **22**, 369–378.

Finch G. R., Haas C. N., Oppenheimer J. A., Gordon G. and Trussel R. R. (2001). Design Criteria for inactivation of *Cryptosporidium* by ozone in drinking water. *Ozone: Science and Engineering*, **23**, 259–284.

Gerba C. P. and Haas C. N. (1988). Assessment of risks associated with enteric viruses in contaminated drinking water. *ASTM Special Technical Publication*, **976**, 489–494.

Gibson III, C. J., Haas C. N. and Rose J. B. (1999). Risk assessment of waterborne protozoa: Current status and future trends. *Parasitology*, **117**, 205–212.

Greenwood M., and Yule G. U. (1917). On the statistical interpretation of some bacteriological methods employed in water analysis. *Journal of Hygiene*, **16**, 36–54.

Gujer W. and von Gunten U. (2003). A stochastic model of an ozonation reactor. *Water Research*, **37**, 1667–1677.

Haas C. N. (1983). Estimation of Risk Due to Low doses of Microorganisms: A Comparison of Alternative Methodologies. *American Journal of Epidemiology*, **118**, 573–582.

Haas C. N., Rose J. B., Gerba C., and Regli S. (1993). Risk assessment of virus in drinking water. *Risk Analysis*, **13**(5), 545–552.

Haas C. N., Rose J. B. and Gerba C. P. (1999). *Quantitative Microbial Risk Assessment*. Wiley, New York, USA.

Havelaar A. H. (1994). Application of HACCP to drinking water supply. *Food Control*, **5**(3), 145–152.

Hijnen W. A. M., Beerendonk E. and Medema G. J. (2005a). *Elimination of Microorganisms by Drinking Water Treatment Processes, second edition*. Kiwa NV, Nieuwegein, the Netherlands.

Hijnen W. A. M., Beerendonk E. and Medema G. J. (2005b). *Additional Memo Including Rapid Sand Filtration, GAC Filtration and Direct Filtration to Elimination of Microorganisms by Drinking Water Treatment Processes*. Kiwa NV, Nieuwegein, the Netherlands.

Hrudey S. E. and Hrudey E. J. (2004). Safe drinking water – Lessons learned from recent outbreaks in affluent nations. IWA Publishing, London.

ILSI Risk Science Institute Pathogen Risk Assessment Working Group (1996). A Conceptual Framework to Assess the Risks of Human Disease Following Exposure to Pathogens. *Risk Analysis*, **16**(6), 841–848.

IWA/WHO, (2004). IWA Bonn Charter for Safe Drinking Water. IWA Publishing, London, UK.

LeChevallier M. and Au K. K. (2004). *Water Treatment and Pathogen Control Process: Efficiency in Achieving Safe Drinking Water.* WHO, IWA publishing, London, UK.

MicroRisk (2002). www.microrisk.com (accessed 1 March 2008).

Medema G. J. (2003). *Water Safety Conference Berlin 28–30 April 2003, Integration of Quantitative Microbial Risk Assessment (QMRA) and Risk Management in Water Safety Plans-Report of a Workshop*. Kiwa, Nieuwegein, the Netherlands.

Medema G. J., Loret J. F., Stenström T. A. and Ashbolt N. (editors) (2006). *Quantitative Microbial Risk Assessment in the Water Safety Plan*, report for the European Commission under the Fifth Framework Programme, Theme 4: 'Energy, environment and sustainable development' (contract EVK1CT200200123), Kiwa Water Research, Nieuwegein, the Netherlands.

Payment P., Berte A., Prévost M., Ménard B. and Barbeau B. (2000). Occurrence of pathogenic microorganisms in the Saint Lawrence River (Canada) and comparison of health risks for populations using it as their source of drinking water. *Canadian Journal of Microbiology*, **46**, 565–576.

Regli S., Rose J. B., Haas C. N. and Gerba C. P. (1991). Modeling the Risk from *Giardia* and Viruses In Drinking Water Journal of the American Water Works Association. *Journal of the American Water Works Association-JAWWA*, **83(**11), 76–84.

Rose J. B. Haas C. N. and Regli S. (1991). Risk Assessment and the Control of Waterborne *Giardia*sis. *American Journal of Public Health*, **81**, 709–713.

Rose J. B., Haas C. N. and Gerba C. P. (1993). Waterborne pathogens: assessing health risks. *Health and environment digest*, **7**(3), 1–8.

SDWA (1974). *Safe Drinking Water Act of 1974*. Public Law 93-523, Dec. 16, 1974. Washington, DC: US Government Printing Office. SDWA. 1977.

Smeets P. W. M. H., Dullemont Y. J. and Medema G. J. (2005). *E. coli* as surrogate for *Campylobacter* inactivation at bench-scale and full-scale and high ozone resistance of environmental *E. coli* and *Campylobacter*. *17th IOA conference 22–26 August 2005, Strasbourg, France.*

Smeets P. W. M. H., Van der Helm A. W. C., Dullemont Y. J., Rietveld L. C., Van Dijk J. C. and Medema G. J. (2006). *E-coli* Inactivation by Ozone Under Bench-scale and Full-scale Hydraulic Conditions. *Water Research*, **40**, 3239–3248.

Smeets P. W. M. H., van Dijk J. C., Stanfield G., Rietveld L. C. and Medema G. J. (2007). How can the UK statutory Cryptosporidium monitoring be used for Quantitative Risk Assessment of Cryptosporidium in drinking water? *Journal of Water and Health*, **5**(S1), 107–118.

Smeets P. W. M. H., Dullemont Y. J., Van Gelder P., Van Dijk J. C. and Medema G. J. (2008). Improved methods for modelling drinking water treatment in quantitative microbial risk assessment; a case study of *Campylobacter* reduction by filtration and ozonation. *Journal of Water and Health*, **6**(3), 301–314.

Taubes G. and Mann C. C. (1995). Epidemiology faces its limits. *Science*, **269**(5221), 164–169.

Teunis P. F. M., Havelaar, A. H. and Medema G. J. (1994). *A Literature Survey on the Assessment of Microbiological Risk for Drinking Water.* RIVM Report 734301006, RIVM Bilthoven, the Netherlands.

Teunis P. F. M., Medema G. J., Kruidenier L. and Havelaar A. H. (1997). Assessment of the risk of infection by *Cryptosporidium* or *Giardia* in drinking water from a surface water source. *Water Research*, **31**(6), 1333–1346.

Teunis P. F. M. and Havelaar A. H. (1999). Cryptosporidium *in Drinking Water. Evaluation of the ILSI/RSI Quantitative Risk Assessment Framework*. Report No. 284 550 006 RIVM, Bilthoven, the Netherlands.

Teunis P., Davison A. and Deere D. (2004). *Short-term Fluctuations in Drinking Water Quality and their Significance for Public Health*. World health organization, Geneva, Switzerland.

USEPA (1999). *Guidance Manual Disinfection Profiling and Benchmarking*. United States Environmental Protection Agency, Washington DC, USA.

USEPA (1989). *SWTR, Surface Water Treatment Rule*. United States Environmental Protection Agency, Washington DC, USA.

USEPA (1998). *IESWTR, Interim Enhanced Surface Water Treatment Rule*. United States Environmental Protection Agency, Washington DC, USA.

USEPA (2002). *LT1ESWTR, Long Term First Enhanced Surface Water Treatment Rule*. United States Environmental Protection Agency, Washington DC, USA.

USEPA (2006). *LT2ESWTR, Long Term Second Enhanced Surface Water Treatment Rule*. United States Environmental Protection Agency, Washington DC, USA.

Van der Veer A. J., Schotsman R. M., Martijn B., Schurer R., van Dijk J. C. (2005). Comparative small technical Scale Study on Ozone Disc Diffusers and Ejector System. *17th IOA conference 22–26 August 2005*, Strasbourg, France.

WHO (2001). *Water Quality: Guidelines, Standards and Health*. Edited by Lorna Fewtrell and Jamie Bartram. World Health Organization Published by IWA Publishing, London, UK.

WHO (2004). *Guidelines for Drinking Water Quality (third edition)*. World Health Organization, Geneva, Switzerland.

# Chapter 8

# General discussion

P.W.M.H. Smeets[a,b], L.C. Rietveld[b], J.C. van Dijk[b] and G.J. Medema[a,b]

[a]KWR Watercycle research institute, PO Box 1072, 3430 BB Nieuwegein, the Netherlands
[b]Faculty of Civil Engineering, Delft University of Technology, PO Box 5048, 2600 GA Delft, the Netherlands

## INTRODUCTION

Since 1980 several drinking water guidelines and legislations changed from monitoring water quality targets, such as the absence of indicator organisms, to setting health-based targets. Health-based targets have been set for treatment performance, water quality, infection risk and disability adjusted life years (DALY). Quantitative microbial risk assessment (QMRA) has been applied to set these targets and to assess whether the targets are met. QMRA has been applied in drinking water since 1989 (USEPA, 1989) and is now part of several guidelines and legislations (Anonymous, 2001; NHMRC, 2004; WHO, 2004 and CDW, 2007). Drinking water treatment plays a crucial role in achieving the health-based targets and the assessment of treatment efficacy is an important part of QMRA. This book illustrated how the reduction of pathogens by drinking water treatment could be quantified for QMRA purposes. This chapter discusses these findings in relation to the questions in Chapter 1 and developments in the field of drinking water safety.

## COMBINING INFORMATION IN THE TREATMENT FRAMEWORK

Many types of data can provide information about treatment efficacy. This information needs to be combined for risk assessment. Recent legislations and guidelines promote the use of site specific information to assess drinking water safety (Anonymous, 2001). In terms of quantification this is often limited to site specific source water monitoring to determine the level of faecal contamination (WHO, 2004a; USEPA, 2006 and CDW, 2007). Treatment performance monitoring, such as turbidity after filtration or disinfectant residual monitoring, is used to verify if the process complies, not to quantify the efficacy. In QMRA, available information is used to quantify drinking water safety. The treatment framework in Chapter 1 was developed to combine site specific full-scale data with the body of information on treatment efficacy from the literature. Treatment efficacy reported in literature ranged over several orders of magnitude for most treatment processes, and therefore provided little information for a site specific risk assessment. Information such as details on process design, (on-line) monitoring of process conditions or microbial monitoring, allowed for a more site specific assessment of treatment efficacy. The framework was applied in a case study in Chapter 2 and in Smeets *et al.* (2006). Site specific information improved the accuracy of the treatment assessment in these studies. This study used data that was already available for the assessed systems. However, the treatment framework can be used beforehand to determine which monitoring efforts would provide the best site specific data at different stages in treatment. Microbial monitoring at the first stages of treatment would provide the most direct, site specific information, however, this would not be feasible at later stages. Full-scale monitoring of process conditions could still provide site specific information about the later stages in treatment.

## INCLUDING VARIABILITY AND UNCERTAINTY BY STOCHASTIC MODELLING

The information collected through the treatment framework will generally consist of discrete values. However, these values actually represent a measurement or analysis of a very small proportion of the total produced water volume (generally less than 0.001% for microbial samples). In addition the measurement or analysis itself includes inaccuracy and uncertainty. Statistical methods can be used to determine what information the reported values actually provide about the level, variability and uncertainty of the measured parameter in the total produced water. Since the concentration of pathogenic organisms in drinking water was generally below detection limit, their concentration was calculated from the concentration in raw water and reduction by treatment. Variability and uncertainty were included in stochastic models of treatment efficacy by describing each step by a probability density function (PDF). These stochastic treatment models, described by Teunis *et al.* (1994, 1997, 1999), were adapted to include different types of data in Chapter 2. Available microbial data was used to determine PDF parameters that described variability of a treatment step. Alternatively, the PDF of treatment efficacy was derived from a range of log-reductions described in literature to determine the uncertainty of using log credits for site specific treatment assessment. The distribution of the pathogen concentration in the treated water was then calculated through Monte Carlo simulation. The stochastic model approach was compared to the more common method of point estimates in case studies in Chapters 2 and 7. Stochastic modelling of treatment showed that, due to the uncertainty and variability of treatment efficacy, treatment was expected to be less effective than the average point estimate. As a consequence the calculated concentration in drinking water and thus the stochastically assessed risk of infection was higher. The stochastic approach showed that the uncertainty of the point estimate of treatment efficacy could range over several orders of magnitude. So point estimates provided a false sense of accuracy. This study applied the treatment efficacies collected from literature assuming that these represent the variability and uncertainty about the efficacy of a treatment process. Most of the reported studies were performed under laboratory or pilot-scale conditions. Therefore the results may not be directly applicable to full-scale systems. In addition methods and data analysis may vary between studies, resulting in differences of observed efficacy. This book therefore focussed on including site specific data to reduce the uncertainty about treatment efficacy. The stochastic method was further improved in Chapters 5 and 6 to distinguish between variability and uncertainty when using microbial monitoring data.

## MICROBIAL MONITORING OF DRINKING WATER

Microbial monitoring of pathogens in drinking water has been used as a basis for QMRA (Lee *et al.*, 2002; Wyn-Jones *et al.*, 2002; Vivier *et al.*, 2002; Mena *et al.*, 2003; Aboytes *et al.*, 2004 and Smeets *et al.*, 2007). However, monitoring of actual pathogens has been limited by the availability of analysis methods, the costs of these methods, method recovery and interpretation of the results. Furthermore, microbial monitoring to assess compliance with a level $10^{-4}$ infection per person per year is not feasible, since this corresponds to extremely low pathogen concentrations. The UK statutory drinking water monitoring for *Cryptosporidium*, discussed in Chapter 3 has provided a direct insight in the way pathogens can be distributed in drinking water. The mean risk of infection is directly related to the mean pathogen concentration when pathogen concentrations are low. Correct assessment of the mean concentration of the (over-dispersed) pathogens in treated water requires sufficient microbial monitoring. Figure 8.1a shows a typical dataset of monitored *Cryptosporidium* concentrations in drinking water. The dashed line in Figure 8.1a indicates combinations of pathogen concentrations and proportions that result in identical mean concentrations. The point where the dashed line touches the monitored concentrations therefore dominated the mean concentration. Both lower and higher monitored concentrations had little impact on the mean concentration since they are further from the dashed line. The mean concentration could be estimated accurately for this dataset since the dominant concentrations were accurately determined. The monitored concentrations resulted from the normal variations of the system, thus providing the continuous smooth curve in Figure 8.1a.

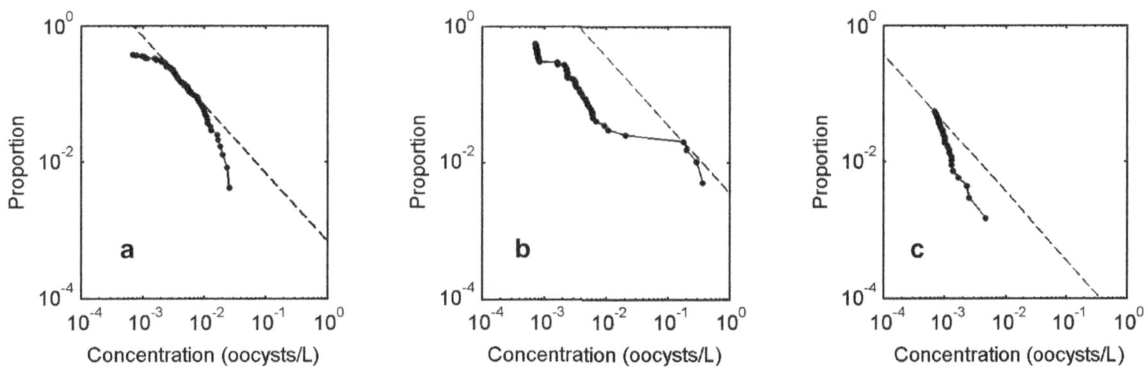


**Figure 8.1** Monitored *Cryptosporidium* concentrations in drinking water (black markers) and combinations of concentrations and proportions that result in identical mean concentrations (dashed line). Figure 8.1a shows a system with normal variations, Figure 8.1b shows a system where special events occurred and Figure 8.1c shows a system with good water quality where the risk is dominated by pathogen concentrations below detection limit

However, at some sites a curve break similar to Figure 8.1b was observed. A curve break could be caused by special events, such as operational error or equipment failure. The dashed line touched the highest concentration of the curve break. Therefore the special event dominated the mean concentration and thus dominated the mean risk. The required microbial monitoring frequency to detect such a special event would be very high, and therefore not practical for routine monitoring. Chapter 7 illustrated how (on-line) process monitoring (e.g. turbidity or disinfectant residual) could be used to detect special events.

Figure 8.1c illustrates the effect of sample volume in a system with good water quality. The dashed line crosses the monitored concentrations at the lowest observed concentration. Therefore even lower concentrations, below detection limit, dominated the mean concentration. Estimating the arithmetic mean concentration with the few positive samples and many negative samples would have resulted in underestimation of the mean concentration. Estimation of the distribution of concentration below the detection limit would dominate the estimate of the mean concentration. The stochastic modelling methods for QMRA presented in this book could be used to estimate how concentrations below the detection limit are distributed, based on the concentrations in source water and reduction by treatment. Thus underestimation of risk could be prevented.

The tree types of monitoring graphs (normal variation Figure 8.1a, special events Figure 8.1b and all or mostly non-detects Figure 8.1c) have all been observed at the 216 UK *Cryptosporidium* monitoring sites. The 'dominant concentration' method described in Chapter 6 provides a tool to determine whether sufficient data was collected for an accurate estimate of the mean concentration. The 'dominant concentration' method also provides guidance for a monitoring plan, supporting decisions on sample volume and frequency. If the dashed line touches at the lowest monitored concentrations, larger sample volumes would provide the most relevant information. If the dashed line touches at the highest monitored concentrations, a higher sampling frequency (possibly with smaller sample volumes) or surrogate monitoring would be required to better quantify normal and special event concentrations. The presented stochastic methods can address the uncertainty of the estimated pathogen concentration in drinking water to support decisions on additional monitoring.

The *Cryptosporidium* concentration in drinking water was compared to the concentration in raw water at seven systems. *Cryptosporidium* removal ranged from 1.8 to 5.2 log, even though all seven systems consisted of coagulation, sedimentation, rapid sand filtration and GAC filtration. This showed that log credits based on unit processes do not account for differences between systems.

## PROCESS MODELLING FOR QMRA

Process models estimate treatment efficacy based on process conditions. Chapters 2 and 7 provided examples of how process models for disinfection could be used in the treatment framework to assess treatment efficacy for QMRA. Disinfection efficacy was estimated based on site specific process conditions. This approach provided an estimate of both the level of inactivation and the variability of the process. Chapter 7 provided examples of process model applications in a QMRA framework. Examples included applications for water safety plans (WSP) to design monitoring, set critical limits, prepare corrective actions and support decisions on optimisation of disinfection.

Chapter 4 demonstrated that the process model had to reflect conditions at full-scale and that proper validation for full-scale environmental conditions was required. Several types of disinfection models ranging from a simple CT calculation to complex Computational Fluid Dynamics (CDF) are being applied today to set process conditions to meet health-based targets (USEPA, 2006 and CDW, 2007). Full-scale validation of an ozonation process in Chapter 4 showed that CT or CT10 models were ineffective to determine if a high level of inactivation (>2 log) was achieved in full-scale systems. Even when residence time distribution was determined by tracer tests, CT and CT10 models did not account sufficiently for hydraulic shortcomings at full-scale. The Continuously Stirred Tank Reactor Model (CSTR) provided an approximation of full-scale hydraulics in the case study. Apart from the choice of hydraulic model type, Chapter 4 showed that the calibration of the model parameters for full-scale environmental conditions was crucial. The most important calibration parameter for inactivation models was the inactivation rate constant of the organism. Experiments showed that this parameter could vary substantially between similar organisms such as *E. coli* and *Campylobacter* and between different strains of one organism (Smeets *et al.*, 2005). In addition the inactivation rate constant varied within the population of a single strain as a function of organism age or level of environmental stress. As a consequence environmental *E. coli* populations had significantly lower inactivation rates than cultured *E. coli*. The cultured *E. coli* were generally used to determine inactivation rates in laboratory studies (Smeets *et al.*, 2005). The findings suggest that conservative inactivation rates should be applied for full-scale modelling. Studies on inactivation rates of pathogens should be conducted under environmental conditions with environmental strains. Conventional applications of disinfection models often lead to more than 10 log overestimation of inactivation. One must take these shortcomings of process modelling into account when models are applied in QMRA and preferably validate the model predictions at full-scale.

Process models for physical treatment, such as coagulation-sedimentation-filtration, have also been reported in literature (Yao, 1971). However calibration of these models is generally complex due to the number of parameters that are difficult to determine in a full-scale system. When these processes are located at the first stages of treatment, microbial monitoring at full-scale can be effective to assess the level and variability of treatment efficacy. Stochastic data analysis and modelling as presented in Chapters 5 and 6 can then be applied to assess treatment efficacy and to predict normal events of poor efficacy. When physical treatment processes are applied at the final stages of treatment, such as slow sand filtration, or when filtration processes are very effective, such as soil passage (artificial recharge, bank filtration, dune infiltration) microbial monitoring may not be effective due to a lack of positives before or after treatment. These processes have been modelled in a QMRA setting (Schijven and Simůnek, 2002 and Van der Wielen *et al.*, 2006) using site specific calibration experiments. Soil, water and microorganism characteristics can have a strong impact on the treatment efficacy and their combined effect in full-scale systems has not been well documented. The lack of positive microbial monitoring data is a general 'problem' for QMRA in the final stages of treatment since safe drinking water contains little microorganisms. Therefore process models for physical processes that calculate treatment efficacy based on measurable parameters need to be developed. These models could improve QMRA studies of effective treatment systems since they could include site specific variability.

Future application of process models in a QMRA framework could be the on-line control of treatment efficacy as part of the SCADA system. The achieved treatment efficacy would become the target parameter instead of individual process parameters.

## QUANTIFYING TREATMENT EFFICACY USING MICROBIAL MONITORING

Microbial monitoring can provide the most direct information about treatment efficacy. Many water utilities have monitored their source water as well as their treated water for faecal indicator organisms. Statistical analysis of such historical data in Chapters 3, 4, 5 and 6 provided insight in the amount and variability of faecal contamination of source and treated waters. The statistical methods described in Chapters 5 and 6 were able to distinguish between variability and uncertainty of the monitored concentrations. Examples showed how microbial monitoring issues such as sample volume, number of samples, non-detects and microbial method (MPN, direct count) could effectively be addressed. Other issues such as microbial method recovery, fraction viable non culturable cells and fraction human pathogenic strains were not demonstrated These could also be addressed through the hierarchical bootstrapping method described in Chapter 5.

Microbial monitoring data before and after a treatment step has been used in several studies to assess the variability of treatment efficacy (Teunis *et al.*, 1997 and Medema *et al.*, 1999). These studies paired the influent and effluent data by date to determine the variability of treatment efficacy. Thus correlation in time between influent and effluent samples was assumed. Chapters 5 and 6 showed that randomly pairing influent and effluent samples resulted in the same assessment of the variability of treatment efficacy as pairing by date. This indicated that the assumed correlation in time was not valid. Alternatively, the rank method paired influent and effluent concentrations by ranked concentration. The rank method assumed that due to the over dispersion of microorganisms in water and the variability of treatment, samples could not be correlated by date. Pairing samples by rank led to an assessment of minimal treatment variability. The treatment efficacies assessed with the date and the rank methods were applied as 'calibration' of the stochastic treatment model. The calibrated model was validated by comparing the predicted concentrations to the monitored concentrations after treatment. In general, pairing by date overestimated variation of treatment efficacy which resulted in an overestimation of concentrations after treatment. The rank method provided a better prediction of concentrations after treatment although it sometimes underestimated concentrations after treatment. So both assumptions of correlation in time and correlation by ranked concentration were not valid.

Therefore an optimisation method was developed in Chapter 6 to calibrate treatment efficacy in the stochastic treatment model by determining the optimal parameters of the treatment PDF. The predicted distribution of concentration after treatment corresponded to the measured concentration when these parameters were used. The calibrated PDF now provided information on the level and variation of treatment efficacy. Thus the calibration of the model could also be considered as the treatment assessment. Presentation of the data in Complementary Cumulative Distribution Function (CCDF) plots allowed visual confirmation of the accuracy of the model. To prevent using the same data for calibration and validation, the model was calibrated and validated by splitting datasets.

## ACCURACY OF STOCHASTIC TREATMENT MODELLING

Stochastic treatment modelling has been applied to predict pathogen concentrations after treatment (Teunis *et al.*, 1997, 1999; Westrell *et al.*, 2003 and Masago *et al.*, 2004). However, the accuracy of these models was not determined. These studies assumed that the stochastic model could be used to predict effluent concentrations including (normal) events. The validity of this assumption was assessed in Chapter 6 by splitting up datasets and using only a part of the dataset to calibrate the stochastic treatment model. Splitting up the data by year and calibrating the model for each year showed that observed treatment efficacy could vary between years. This observation could be caused by the limitations of the monitoring (e.g. limited number of samples, overdispersion of microorganisms in water). The calibrated model for five years was used to predict concentrations after treatment for individual years over a ten-year period. The predicted concentrations of the years were compared to the monitored concentrations to assess the validity of the model. The calibrated model did predict the level and variation of microorganism concentrations after treatment, even when these levels varied between years. So the applicability of stochastic treatment modelling in this case study was verified.

Many QMRA studies had to use surrogate organism monitoring to calibrate the stochastic treatment model for pathogens (Teunis *et al.*, 1999 and Medema *et al.*, 1999). Surrogate organisms for pathogens were chosen based on similarities in organism characteristics and behaviour during treatment. Common surrogate-pathogen combinations were SSRC-*Cryptosporidium* for coagulation, filtration and disinfection (Teunis *et al.*, 1997 and Hijnen *et al.*, 2002), *E. coli*-*Campylobacter* for coagulation, filtration and disinfection (Smeets *et al.*, 2005) and bacteriophages (MS2, PRD1)-enteric viruses for filtration and transport through soil (Schijven and Simůnek,

2002). The validity of these surrogates was generally assessed in pilot tests. However, data analysis at full-scale often indicated poor correlation between pathogen and surrogate reduction by treatment (Chapter 6, Smeets *et al.*, 2005). Chapter 6 showed that surrogate data could be used to effectively calibrate a treatment model for pathogens, even when pairing pathogen and surrogate data by date indicated poor correlation. Due to the complexity and costs involved in pathogen monitoring, treatment validation through surrogate organisms will continue to play an important role in the future.

## APPLICATIONS OF QMRA IN THE WSP

QMRA in itself does not improve drinking water safety. Chapter 1 discussed the use of QMRA as a tool in Water Safety Plans (WSP). Several examples of QMRA applications in the WSP were demonstrated in Chapter 7.

### System assessment

Assessment of the system's capability to reach health-based targets is part of the WSP. The system assessment in the WSP (WHO, 2006) uses a risk matrix for a semi-quantitative risk assessment. Chapters 2 to 6 provided examples of how treatment efficacy can be assessed quantitatively and objectively. These examples made clear that point estimates using log credits and CT10 models are not sufficient for site specific assessment of treatment efficacy. Risk managers may overlook the impact of 'normal' events during semi-quantitative assessments due to the fact that they occur as part of normal variations. The presented stochastic methods were especially effective to quantify the frequency and magnitude of 'normal' events that result from regular variations in the system.

Some events are the result of operational errors or equipment failure. These were referred to as 'special' events. Risk managers may estimate the frequency and magnitude of special events based on operational experience and failure reports or on other methodologies, such as Failure Mode Analysis (FMA). QMRA can quantify the effect of special events on risk to support decisions in the WSP. Examples of how these special events can be incorporated in QMRA were provided in Chapter 7.

### Risk prioritisation

Risk prioritisation can be performed at different levels of detail. A point-estimate QMRA using log credits can be sufficient to identify sites with the highest risk. When prioritising risk at a specific site, the risk estimate generally needs to be more accurate. The example of chlorine disinfection optimisation in Chapter 7 showed the relative priorities of average performance versus chlorine failure events. Improving the hydraulics of a chlorine disinfection system would significantly increase nominal inactivation. However, yearly mean inactivation would only improve if rare events of low chlorine residual were also eliminated. Both hydraulic adaptation and elimination of chlorine failure were more effective than increasing the chlorine dose.

### Design monitoring

Monitoring of drinking water treatment serves to assess water quality and treatment efficacy and to verify that no significant failure events occur. Microbial monitoring provided the most accurate site specific information of full-scale treatment performance. The dominant concentration concept presented in Chapter 6 can be used to design microbial monitoring programmes. Based on monitoring results, the method assessed whether monitoring was sufficient to accurately assess the mean concentration (Figure 8.1a). When the mean concentration cannot be determined accurately from the data, the method provides guidance on monitoring strategies. The strategy could consist of frequent sampling of small volumes to quantify the frequency and magnitude of normal events. Or it could consist of less frequent sampling of larger volumes to quantify nominal concentrations (Figure 8.1c). In some cases larger sample volumes are not feasible. In that case stochastic treatment modelling can be used to estimate these concentrations based on concentrations before treatment and the assessed treatment efficacy.

The required monitoring frequency to detect relevant failure events was discussed in Chapter 7. On-line monitoring of parameters that would indicate an event, such as turbidity, chlorine residual, flow or equipment monitoring, could be used since detecting the event is more important than quantifying efficacy. As treatment efficacy increased, more frequent monitoring was required to detect significant events. Daily monitoring could only verify two log reduction. Monitoring every ten seconds was required to verify six log reduction by a single treatment step. A multiple barrier system providing six log reduction would require less frequent monitoring since three processes each providing two log reduction would only require daily monitoring of each treatment step.

## Setting critical limits

A treatment system can be designed to provide exactly the right level of treatment to meet the health-based targets. However, in practice the risk manager needs to account for variations and inaccuracies in order to run a practical and stable process. Setpoints for processes such as disinfectant dose or filter run time are set accordingly and may be adapted for changes such as flow or temperature. Critical limits must be set to operational or monitoring parameters. As long as these limits are not exceeded, the produced water will comply with the health-based target. Examples in Chapter 7 showed that even a short deviation might lead to not reaching the yearly average health target. Therefore setpoints and operational limits must provide some safety margin that allows for periodic deviation and correction. The required margin depends on the level of monitoring, the response time needed to bring a setpoint back on specification and the expected frequency and magnitude of events. An example in Chapter 7 showed that a chlorine disinfection system required a setpoint corresponding to 4 log reduction to achieve 3 log reduction. On a yearly basis this would allow for an eight-hour failure to occur once a year. In that case the critical limit would equal the yearly target of 3 log reduction.

Ideally the processes could be controlled on-line to achieve a level of pathogen reduction, rather than applying setpoints and critical limits for individual process parameters. Thus changes in flow or process conditions could be adequately addressed. In multiple barrier systems, the combined effect of the barriers, calculated with QMRA, could be controlled on-line. Deviance of one barrier could then be mitigated by the margin or adaptation of other barriers. Thus robust, effective and efficient operation could be achieved.

## Corrective actions

When critical limits are exceeded to the extend that normal process control cannot bring the system back to specification, corrective actions are required. QMRA was used in Chapter 7 to determine the allowable response time in relation to the level of the corrective action. In general, corrective actions need to be taken rapidly. An example showed that adequate action was required within 6.5 hours in order to prevent exceeding the yearly treatment disinfection target of 2.5 log inactivation, when loss of disinfectant residual was detected. The QMRA studies of full-scale data in Chapters 3, 5 and 6 showed that risks were generally dominated by normal events occurring between 1% and 10% of the time. They also showed that special events occurring less than 0.1% of the time could also dominate the mean risk.

## IMPLICATIONS FOR THE DRINKING WATER INDUSTRY

Water utilities have the responsibility to provide safe drinking water. Verifying that drinking water treatment effectively removes or inactivates pathogens from the source water is a crucial part of this responsibility. This study showed that point estimates based on literature can overestimate treatment efficacy. Point estimates using site specific monitoring data improve the estimate, but can still overestimate efficacy. Stochastic models include variability of water quality and treatment and therefore provide a better estimate of pathogen concentration in drinking water. However, the currently applied method of pairing microbial samples before and after treatment by date often leads to underestimation of treatment efficacy and thus to an overestimation of risk. The methods in this book have overcome this problem and can provide a more reliable estimate of risk. In addition they provide insight in the variability of risk and the uncertainty of the assessment.

The stochastic methods are more complicated than point estimates and may claim more resources to collect data. The benefits of stochastic method can outweigh these disadvantages. Point estimates can also underestimate treatment efficacy leading to an overestimation of risk and collecting more data to reduce uncertainty can be more efficient than expanding treatment.

Limited microbial monitoring can lead to overestimation of pathogen concentrations during events and thus to overestimation of the yearly average risk of infection. Measures to reduce this risk, such as additional treatment, will likely require more resources than extra monitoring. Chapter 6 provided guidance on monitoring strategies, such as sample volume and frequency to determine when monitoring has been 'enough'.

Currently monitoring often leads to 'data graveyards' without any value. By directing monitoring efforts towards use for QMRA this data would become valuable. Apart from monitoring design (sampling location, frequency and volume), this should include choice of microbial method (how quantitative is the result?) and data collection and storage (store actual counts and sample volumes, microbial method used, uniform parameters names etc.). QMRA methods could be used beforehand to address some of these issues and provide an effective and efficient monitoring programme that would lead to statistically valid conclusions. The overview of 20 case studies of full-scale efficacy in Chapter 6 can be used by water utilities as a benchmark of what other systems have achieved in practice.

The frequency and magnitude of 'special' events (e.g. due to equipment failure) cannot be predicted with the stochastic methods. Therefore water safety plans (WSP) are needed to identify 'special' events. Chapter 8 provided several examples of how QMRA can be used as a tool in the WSP. In many cases this included process modelling. Water utilities need to realise that a full-scale treatment plant is not a large laboratory. Chapter 4 showed that process models that are effective to describe laboratory experiments can overestimate treatment efficacy by many orders of magnitude when they are directly applied to full-scale situations. This is caused by hydraulic shortcomings, variation of conditions and long term effects at full-scale and the differences between cultured and environmental microbial populations. Site specific validation of process models at full-scale is therefore required, especially if a process is expected to achieve high pathogen reduction exceeding 2 logs.

Health-based targets are generally set over a longer period of time, such as yearly risk of infection. This could lead to the assumption that a short period of non-compliance is acceptable since the risk is averaged out. Since treatment systems generally need to achieve several log units of pathogen reduction, even a few hours of non compliance could lead to not meeting the yearly health-based target. Water utilities need to train their operators to realise the importance of continuous effective treatment. As an example, treatment should never be compromised for operational changes (e.g. start up of a treatment line) or maintenance (e.g. chlorine pump or ozone generator maintenance).

## CONSIDERATIONS FOR THE REGULATORS

Recently QMRA has been incorporated in several guidelines and legislations, each with a different approach and limit (Anonymous, 2001; NHMRC, 2004; WHO, 2004a and CDW, 2007). Guidelines have applied either the risk of infection or disability adjusted life years (DALYs) to define health-based targets (WHO, 2004a). The Dutch drinking water regulations (Anonymous, 2001) require water companies to demonstrate compliance with a $10^{-4}$ yearly risk of infection target by a site specific QMRA. In Canada the Federal-Provincial-Territorial Committee on Drinking Water (CDW) is using QMRA in the development of new 'Guideline Technical Documents for both Enteric Virus and Protozoa' (CDW, 2007). The draft guideline proposes 4 log reduction of viruses and 3 log removal of protozoa in combination with a health-based target of $10^{-6}$ or $10^{-7}$ DALY. The new Australian drinking water guidelines have incorporated QMRA, although it is recognised that generally not enough data will be available (NHMRC, 2004). In the USA the LT2ESWTR was released (USEPA, 2006). Although the preparations of the initial SWTR in 1994 were somewhat guided by the target of $10^{-4}$ risk of infection from *Giardia*, it is not the target of the LT2ESWTR to achieve a quantified level of safety (Ashbolt, 2007).

Remarkably most legislations allow the use of log credits and CT10 models to quantify treatment efficacy. Log credits can be effective for an initial prioritisation of sites at risk. However, in a site specific assessment log credits can both over and underestimate treatment efficacy by several orders of magnitude (Chapters 2, 3, 4 and 7). As a consequence this approach could lead to unnecessary investments or a false sense of safety. Although legislations and guidelines recognised the variability of risk and the uncertainty of the QMRA results, no clear targets for acceptable events or the required level of uncertainty were set.

With respect to variability, high risk events might be of concern since these could lead to an outbreak situation. This leads to the question whether the health-based target of $10^{-4}$ risk of infection per person per year sufficiently covers outbreak conditions. Incidence of gastroenteritis in affluent countries range from 0.1 to 3.5 episodes per person per year (Roy *et al.*, 2006) or 1%–2% incidence of Cryptosporidiosis (Casman *et al.*, 2000 and Van Pelt and Van Duynhoven, 2006). Therefore the nominal $10^{-4}$ risk of infection from drinking water would contribute less than 10% to the endemic level of gastroenteritis, which seems appropriate. However, short events of high risk could lead to an outbreak situation resulting in social concern which can be referred to as 'risk aversion' (similar to airplane crashes versus car crashes (Vrijling *et al.*, 1998)). Therefore Signor *et al.* (2007) promoted the use of a daily risk target of $10^{-6}$ risk of infection per person per day (the yearly $10^{-4}$ health target corresponds to an average daily risk of infection of $2.74*10^{-7}$ per person per day). With a yearly target of $10^{-4}$, theoretically a daily risk of $10^{-4}$ would be acceptable one day per year when the risk on all other days would be negligible (e.g. $<10^{-9}$). If on this day each infection would lead to illness, this would lead to 0.01% of the population becoming ill. This is far below the threshold of detection of an outbreak through health monitoring which is around 1% (Regli *et al.*, 1991). Therefore the $10^{-4}$ health target is sufficient both for average risk and risk aversion. Legislators and risk managers need to be aware however that this average health target can be dominated by short rare events of high risk, and therefore quantification of these events in QMRA is crucial.

The health target is presented as a discrete value, however in risk assessment some level of uncertainty around this value arises. Choosing an appropriate level of certainty is a political and social point of discussion rather than a technical one. The upper 95% confidence level (CL) is a level of certainty that is often used. Choosing the 50%, 90%, 95% or 99% confidence level may have an impact of several orders of magnitude with regard to the required treatment. For example, in Chapter 7, the median log reduction of *Cryptosporidium* by filtration reported in literature was 2 log (50% CL). However, when all the reported log reductions were incorporated in a stochastic assessment, the expected reduction was 1.5 log. The 90%, 95% and 99% CL were 1.1, 0.8 and

0.3 log reduction respectively, so the estimate of filtration efficacy could vary two orders of magnitude depending on the chosen confidence level. The stochastic method in Chapter 5 distinguished between the variability of risk and the uncertainty of the assessment, resulting in the FN-curve. This would allow legislators to decide whether both the average and the peak risk were acceptable and whether this was sufficiently certain.

The application of QMRA in drinking water is still being developed, and water utilities need to make some crucial choices that are not site specific and are not in the field of expertise of water suppliers. These include:

- the choice of QMRA method;
- the selection of pathogens;
- dose-response relations;
- health effects risks and severity weight (DALY).

Since these choices could significantly affect the QMRA outcome, legislators should provide guidance for a uniform approach when QMRA is applied for legal compliance.

## Choice of QMRA method

The OMRA method strives to provide an objective estimate of the risk of infection through statistical methods. Stochastic modelling of drinking water treatment through Monte Carlo simulation has become an accepted method for QMRA applications (Haas *et al.*, 1999). However, slightly different methods for site specific stochastic treatment model calibration were reported in literature (Teunis *et al.*, 1997, 1999; Petterson *et al.*, 2006 and Smeets *et al.*, 2007). The choice of method can have a significant impact on the assessed risk. The methods proposed in this book require few prior assumptions and allow intuitive, visual evaluation of the modelling results and comparison with monitoring results. The current challenge to the QMRA experts is to provide a unified QMRA approach for the water industry and legislators.

## Selecting pathogens

Recently new types and strains of human pathogenic microorganisms were identified as possibly being relevant for drinking water safety. Evidence of transitions between hosts due to genetic adaptation has raised concern for new viruses such as SARS (Kuiken *et al.*, 2003) and Avian influenza (De Jong and Hien, 2006). Another group of pathogens that is receiving increased attentions are zoonotic parasites, such as *Cryptosporidium*, *Giardia*, *Toxoplasma*, *Cyclospora cayetanensis*, Amebae and ciliated protozoa, *Blastocystis*, microsporidia and zoonotic helminths (Various, 2004a and WHO, 2004b). But also known microbial parameters that were previously of less hygienic concern receive renewed attention, such as HPC infections in immunocompromised (Various, 2004b; Glasmacher *et al.*, 2003). Bioterrorist infectious agents could also include new threats to drinking water (Todd, 2006 and CAMRA, 2007). It will not be feasible to asses the risk of infection for all these individual organisms. Therefore it is suggested to assesses the risk for a suite of index pathogens that are expected to cover the challenges posed by these existing and new microbes (MicroRisk, 2006). This book showed that the limited knowledge about the index pathogens in drinking water caused uncertainties in QMRA. Improving our knowledge on the occurrence of index pathogens in source water, reduction by treatment and health consequences should be the focus of research. By relating this knowledge to characteristics of emerging pathogens the risk from these pathogens can be assessed.

## Dose-response

The impact of choice of dose-response is illustrated by two studies. Masago *et al.* (2004) found that absence of *Cryptosporidium* in daily samples of 180 L was sufficient to verify that the $10^{-4}$ health target was met, however Smeets *et al.* (Chapter 3) found that daily sampling of 1,000 L was not sufficient. These differences were caused by the applied dose-response models (Haas *et al.*, 1996 respectively Teunis *et al.*, 2002). In addition, susceptibility of the population may change in time due to demographic changes, such as aging of the population, increase of immunocompromised people due to healthcare, decrease of general immunity due to reduced exposure to pathogens and increase of infectivity due to changes in the genetic structure of viruses. To overcome the uncertainty about the dose-response the maximum risk curve could be used to provide a maximum estimate.

### Health effect and severity weight

To determine the DALY additional assumptions about the health outcome risk following infection and the severity weight of this outcome need to be determined. These are relevant when different hazards are compared, e.g. disinfection by-products versus risk of infection (Havelaar *et al.*, 200 and Ashbolt, 2004). However, when setting targets for individual hazards, the DALY approach introduces significant uncertainties. Choices made to calculate the DALY could dominate the outcome of the assessment and should therefore be made explicitly clear when reporting the results.

### Guidance by the legislator

Water utilities have the expertise to make decisions with respect to source water, treatment and distribution of drinking water. Therefore legislation and guidance for water utilities should be directed at these issues that can be controlled by the water utilities. The water utilities will be capable of estimating the site specific dose distribution within the supplied population. Legislators should provide guidance to translate this dose to the resulting health risk targets.

## FUTURE RESEARCH

Although the knowledge on QMRA has progressed over the last two decades, the drinking water industry is still just at the beginning of applying QMRA on a regular (regulatory) basis. This leads to new insights and adaptations of methods. Research on the following issues can reduce the uncertainty in treatment assessment and enable a broader implementation of QMRA, thus leading to safer drinking water.

### Process models

Physical treatment processes such as coagulation, sedimentation, filtration and soil passage form important barriers of our drinking water systems. Microbial monitoring is not effective to assess the efficacy of these treatment processes when they are applied at the final stages of drinking water treatment. Process models that can estimate the efficacy of these processes based on measurable parameters could therefore improve the risk assessment for these final stages of treatment. Small scale drinking water systems and systems in developing countries generally do not have the resources for extensive microbial assessment. Process models could provide a relatively cheap and fast tool to assess and improve the safety of these systems.

Some of the relevant parameters for physical treatment, such as size of the microorganism and its surface charge have been studied under laboratory condition. However, little is known about the state of microorganisms under environmental full-scale conditions. Microorganisms may be agglomerated or attached to particles which changes their size and surface characteristics. The characteristics will not be identical for all microorganisms, filter material and water conditions and the efficacy of a process will be determined by the proportion of the system with unfavourable characteristics for pathogen removal. Future research should focus on the development of process models that use measurable parameters to determine these characteristics.

The challenge of this research will be to effectively address complications of environmental and full-scale conditions. Therefore sufficient information from full-scale systems is needed to provide feedback on the applicability of the process models. This book has provided a limited overview of full-scale treatment efficacy. The overview showed that in general processes are less effective at full-scale than at pilot scale. However, individual systems were less variable than would be expected based on experimental findings. The overview of full-scale treatment efficacy could provide initial feedback for the process models. This could be expanded by collecting data from more full-scale systems and including the model parameters in the process monitoring.

## Indexing new pathogens

As new pathogens emerge their risk for infection through drinking water needs to be assessed. Based on their general characteristics such as occurrence in the environment, pathogenicity and known outbreaks though water, the pathogen may be indexed as a potential hazard for drinking water. The efficacy of drinking water treatment for this new pathogen needs to be assessed. It is impractical to perform experiments for each new pathogen and treatment process. Future research needs to determine what characteristics make an organism susceptible to or resistant against a treatment process. This could relate to the research on physical processes, where surface properties, size, shape, mobility and persistence determine the efficacy. Other properties, such as membrane composition and capsules may be relevant for disinfection processes, whereas DNA or RNA structure and repair mechanisms will impact the effectiveness of UV treatment. Emerging pathogens can then be indexed based on these characteristics without the need for extensive experiments with each organism.

## Interaction between processes

The stochastic treatment model assumed independence between treatment steps. However, some interaction between processes is expected, e.g. failure of particle removal may reduce the efficacy of consecutive disinfection processes. Such common cause failures could lead to an event. Therefore interaction between processes needs to be studied in detail. Situations where this interaction occurs could then be modelled as special events in the stochastic model.

## REFERENCES

Anonymous (2001). Adaptation of Dutch drinking water legislation. Besluit van 9 januari 2001 tot wijziging van het waterleidingbesluit in verband met de richtlijn betreffende de kwaliteit van voor menselijke consumptie bestemd water. *Staatsblad van het Koninkrijk der Nederlanden*, **31**, 1–53.

Ashbolt N. J. (2004). Risk analysis of drinking water microbial contamination versus disinfection by-products (DBPs). *Toxicology*, **198**, 255–262.

Ashbolt N. J. (2007). Personal Communication. USEPA, Cincinnati, USA.

Aboytes R., Di Giovanni G. D., Abrams F. A. and Rheinecker C. (2004). Detection of Infectious. *Cryptosporidium* in filtered drinking water. *Journal American Water Works Association*, **96**(9), 88–98.

Casman E. A., Fischhoff B., Palmgren C., Small M. J. and Wu F. (2000). An integrated risk model of a drinking-water-borne cryptosporidiosis outbreak. *Risk Analysis*, **20**(4), 495–511.

CAMRA (2007). Center for Advancing Microbial Risk Assessment. http://www.camra.msu.edu (accessed 1 March 2008).

CDW (2007). *Approach for CDW Discussion and Decision; Microbiological Documents – Enteric Viruses and Protozoa* Draft Discussion Document – ITEM 3.04 October 2007. Federal-Provincial-Territorial Committee on Drinking Water (CDW) Health Canada, Ottawa (Ontario) Canada.

Craig L. (2004). Conference statement on priorities or drinking-water safety in Canada. *Journal of Toxicology and Environmental Health*, Part A, **67**(20), 1551–1553.

Davison A., Deere D., Stevens M., Howard, G. and Bartram J. (2006). *Water Safety Plan Manual*, WHO, Geneva.

De Jong M. and Hien T. (2006). Avian influenza A (H5N1). *Journal of Clinical Virology*, **35**(1), 2–13.

Glasmacher A., Engelhart S. and Exner M. (2003). Infections from HPC organisms in drinking-water amongst the immunocompromised. In: *World Health Organization (WHO). Heterotrophic Plate Counts and Drinking-water Safety*. J. Bartram, J. Cotruvo, M. Exner, C. Fricker, A. Glasmacher (eds.). IWA Publishing, London, UK.

Haas C. N., Crockett C. S., Rose J. B., Gerba C. P. and Fazil A. M. (1996). Assessing the risk posed by oocysts in drinking water. *Journal of the American Water Works Association*, **88**(9), 131–136.

Haas C. N., Rose J. B. and Gerba C. P. (1999). *Quantitative Microbial Risk Assessment*. Wiley, New York, USA.

Havelaar A. H., De Hollander A. E. M., Teunis P. F. M., Evers E. G., Van Kranen H. J., Versteegh J. F. M., Van Koten J. E. M. and Slob W. (2000). Balancing the risks and benefits of drinking water disinfection: Disability adjusted life-years on the scale. *Environmental Health Perspectives*, **108**(4), 315–321.

Howard G., Pedley S. and Tibatemwa S. (2006). Quantitative microbial risk assessment to estimate health risks attributable to water supply: Can the technique be applied in developing countries with limited data? *Journal of Water Health*, **4**(1), 49–65.

Hijnen W. A. M., Van der Veer A. J., Van Beveren J. and Medema G. J. (2002). Spores of sulphite-reducing clostridia (SSRC) as surrogate for verification of the inactivation capacity of full-scale ozonation for *Cryptosporidium. Water science & technology: water supply*, **2**(1), 163–170.

Hrudey S. E. and Hrudey E. J. (2004). *Safe Drinking Water – Lessons Learned from Recent Outbreaks in Affluent Nations*. IWA Publishing, London.

Kuiken T., Fouchier R. A. M., Schutten M., Rimmelzwaan G. F., Van Amerongen G., Van Riel D., Laman J. D., De Jong T., Van Doornum G., Lim W., Ling A. E., Chan P. K. S., Tam J. S., Zambon M. C., Gopal R., Drosten C., Van der Werf S., Escriou N., Manuguerra J. C., Stöhr K., Peiris J. S. M., Osterhaus A. D. M. E. (2003). Newly discovered coronavirus as the primary cause of severe acute respiratory syndrome. *Lancet*, **362**(9280), 263–270.

Lee S. H. and Kim S. J. (2002). Detection of infectious enteroviruses and adenoviruses in tap water in urban areas in Korea. *Water Research*, **36**, 248–256.

Mena K. D., Gerba C. P., Haas C. N. and Rose J. B. (2003). Risk Assessment of Waterborne Coxsackievirus. *Journal of the American Water Works Association-AWWA*, **95**(7), 122–131.

Masago Y., Oguma K., Katayama H., Hirata T. and Ohgaki S. (2004). *Cryptosporidium* monitoring system at a water treatment plant, based on waterborne risk assessment. *Water Science and Technology*, **50**(1), 293–299.

Medema G. J., Groennou J., Hijnen W. A. M., Teunis P., Kruidenier L., Ketelaars H., Willemsen-Zwaagstra J., Havelaar A. and Van der Kooij D. (1999). Frequency distributions of *Cryptosporidium* and *Giardia* in raw water and elimination capacity of water treatment. In: *Cryptosporidium and Giardia: new challenges to the water industry*, Medema, G. J. Thesis, University of Utrecht, the Netherlands.

Microrisk (2007). *Quantitative Microbial Risk Assessment in the Water Safety Plan*, Medema, G., Loret, J., Stenström, T. A. and Ashbolt, N. (eds.), report for the European Commission under the Fifth Framework Programme, Theme 4: 'Energy, environment and sustainable development' (contract EVK1CT200200123), Kiwa Water Research, Nieuwegein, the Netherlands.

NHMRC (2004). *Australian drinking water guidelines 6 Endorsed by 10–11 April 2003*, National Health and Medical Research Council, Australia.

Petterson S., Signor R., Ashbolt N. and Roser D. (2006). *QMRA methodology*. Chapter 7 in *Quantitative Microbial Risk Assessment in the Water Safety Plan*, Medema G., Loret J., Stenström T. A. and Ashbolt N. (eds.), report for the European Commission under the Fifth Framework Programme, Theme 4: 'Energy, environment and sustainable development' (contract EVK1CT200200123), Kiwa Water Research, Nieuwegein, the Netherlands.

Regli S., Rose J. B., Haas C. N. and Gerba C. P. (1991). Modeling the Risk from *Giardia* and Viruses in drinking water. *Journal of the American Water Works Association JAWWA*, **83**(11), 76–84.

Roy S. L., Scallan E. and Beach M. J. (2006). The rate of acute gastrointestinal illness in developed countries. *Journal of Water and Health*, **4**(S2), 31–70.

Schijven J. F. and Simůnek J. (2002). Kinetic modeling of virus transport at the field scale. *Journal of Contaminant Hydrology*, **55**(1–2), 113–35.

Signor R. S. and Ashbolt N. J. (2007). On comparing probabilistic microbial risk assessments against daily rather than annualised acceptable infection probability targets. *14th International Health Related Water Microbiology Symposium,* Tokyo, Japan, 9th–15th September 2007.

Smeets P. W. M. H., Dullemont Y. J. and Medema G. J. (2005). *E. coli* as surrogate for *Campylobacter* inactivation at bench-scale and full-scale and high ozone resistance of environmental *E. coli* and *Campylobacter. 17th IOA conference 22–26 August 2005*, Strasbourg, France.

Smeets P. W. M. H., Hijnen W. A. M. and Stenström T. A. (2006). Efficacy of water treatment processes, Chapter 4 in *Quantitative Microbial Risk Assessment in the Water Safety Plan*, Medema, G., Loret, J., Stenström, T. A. and Ashbolt, N. (eds.), report for the European Commission under the Fifth Framework Programme, Theme 4: 'Energy, environment and sustainable development' (contract EVK1CT200200123), Kiwa Water Research, Nieuwegein, the Netherlands.

Smeets P. W. M. H., Van Dijk J. C., Stanfield, G., Rietveld, L. C. and Medema, G. J. (2007). How can the UK statutory *Cryptosporidium* monitoring be used for Quantitative Risk assessment of *Cryptosporidium* in drinking water? *Journal of Water and Health*, **5**(S1), 107–118.

Teunis P. F. M., Havelaar A. H. and Medema G. J. (1994). *A literature Survey on the Assessment of Microbiological Risk for Drinking Water*. RIVM Report 734301006, RIVM Bilthoven, the Netherlands.

Teunis P. F. M., Medema G. J., Kruidenier L. and Havelaar A. H. (1997). Assessment of the risk of infection by *Cryptosporidium* or *Giardia* in drinking water from a surface water source. *Water Research*, **31**(6), 1333–1346.

Teunis P. F. M., Evers E. G. and Slob W. (1999). Analysis of variable fractions resulting from microbial counts *Quantitative Microbiology*, **1**, 63–88.

Teunis P. F. M. and Havelaar A. H. (1999). Cryptosporidium *in drinking water: Evaluation of the ILSI/RSI quantitative risk assessment framework.* Report 284550006 RIVM, Bilthoven, the Netherlands.

Teunis P. F. M., Chappell C. L. and Okhuysen P. C. (2002). *Cryptosporidium* dose-response studies: variation between isolates. *Risk Analysis*, **22**(1), 175–183.

Todd E. (2006). Building Microbial Risk Assessment to Address Bioterrorism. *Society of Risk Assessment Annual Meeting,* Baltimore, Maryland USA.

USEPA (1989). *SWTR, Surface Water Treatment Rule*. United States Environmental Protection Agency, Washington DC, USA.

USEPA (2006). *LT2ESWTR, Long Term Second Enhanced Surface Water Treatment Rule*. United States Environmental Protection Agency, Washington, DC, USA.

Various (2004a). Zoonoses dedicated issue of Veterinary. *Parasitology*, **126**, 2004.

Various (2004b). HPC dedicated issue of *International Journal of Food Microbiology*, **92**, 239–390.

Van der Wielen P. W. J. J., Blokker M. and Medema G. J. (2006). Modelling the length of microbiological protection zones around phreatic sandy aquifers in the Netherlands. *Water Science & Technology*, **54**(3), 63–69.

Van Lieverloo J. H. M., Blokker M. and Medema G. J. (2007). Quantitative microbial risk assessment of distributed drinking water using faecal indicator incidence and concentrations. *Journal of Water and Health,* **5**(S1), 131–149.

Van Pelt W. and Van Duynhoven Y. (2006). *Trends in Gastro-enteritis in Nederland; notitie met betrekking tot 2005 (Trends in Gastro-enteritis in the Netherlands; memo with regard to 2005)* 04-Oktober-2006, RIVM Bilthoven, the Netherlands.

Vivier J. C., Ehlers M. M., Grabow W. O. K. and Havelaar A. H. (2002). Assessment of the risk of infection associated with Coxsackie B viruses in drinking water. *Water Science and Technology: Water Supply*, **2**(3), 1–8.

Vrijling J. K., Van Hengel W. and Houben R. J. (1998). Acceptable risk as a basis for design. *Reliability Engineering and System Safety*, **59**(1), 141–150.

Westrell T., Bergstedt O., Stenström T. A. and Ashbolt N. J. (2003). A theoretical approach to assess microbial risks due to failures in drinking water systems. *International Journal of Environmental Health Research*, **13**(2), 181–197.

WHO (2004a). *Guidelines for Drinking Water Quality (third edition)*. In: World health organization, Geneva, Switzerland.

WHO (2004b). *Waterborne zoonoses: identification, causes and control*, J. A. Cotruvo, A. Dufour, G. Rees, J. Bartram, R. Carr, D. O. Cliver, G. F. Craun, R. Fayer and V. P. J. Gannon (eds.). IWA Publishing, London.

WHO (2005). *Water Safety Plans Managing Drinking-water Quality from Catchment to Consumer.* World health organization, Geneva, Switzerland.

Wyn-Jones A. P., Watkins J., Francis C., Laverick M. and Sellwood J. (2002). Enteric viruses in drinking water supplies. *Water Science and Technology: Water Supply*, **2**(3), 17–22.

Yao K. M., Habibian M. T. and O'Melia C. R. (1971). Water and wastewater filtration. Concepts and applications. *Environmental Science and Technology*, **5**(11), 1105–1112.

IWA Publishing's authorised EU representative for General Product Safety Regulations is Diane D'Arras, 15 rue Duret, 75116 Paris, France, e-mail: safety@iwap.co.uk.

Printed and bound by CPI Group (UK) Ltd, Croydon, CR0 4YY
12/05/2026
02108870-0019